Shopping
Our Way to Safety

Shopping
Our Way to Safety

How We Changed from
Protecting the Environment
to Protecting Ourselves

Andrew Szasz

University of Minnesota Press   Minneapolis / London

Published by the University of Minnesota Press
111 Third Avenue South, Suite 290
Minneapolis, MN 55401-2520
http://www.upress.umn.edu

Library of Congress Cataloging-in-Publication Data

Szasz, Andrew, 1947–
    Shopping our way to safety : how we changed from protecting the
environment to protecting ourselves / Andrew Szasz.
        p.   cm.
    Includes bibliographical references and index.
    ISBN: 978-0-8166-3508-5 (hc : alk. paper)
    ISBN-10: 0-8166-3508-0 (hc : alk. paper)
    ISBN: 978-0-8166-3509-2 (pb : alk. paper)
    ISBN-10: 0-8166-3509-9 (pb : alk. paper)
    1. Environmental economics—United States.   2. Consumption
(Economics)—United States.   I. Title.
    HC110.E5S93 2007
    339.4'70973—dc22

                                                            2007026438

Printed in the United States of America on acid-free paper

The University of Minnesota is an equal-opportunity educator and employer.

12 11 10 09 08 07   10 9 8 7 6 5 4 3 2 1

*In memory of
my father,
Klein Miklós,
1907–1950*

Contents

Acknowledgments

I started to think about the core idea in this book, the phenomenon I would eventually call *inverted quarantine,* in 1982, when I walked into a supermarket in Highland Park, New Jersey, and came face to face with what seemed to me an entire aisle filled with bottled waters. If dated from that moment, then, it has taken me twenty-five years, more or less, to bring this project to completion. Other work and other life events took precedence for long stretches during those years; I switched jobs, got married, had my first child, wrote another book, wrote articles, got tenure, had two more kids, taught many classes.

At the meetings of the American Sociological Association (ASA) in 1994, I presented some of the core ideas in this book, in a preliminary fashion, in a paper titled "Inverted Quarantine: Bomb Shelters, Bottled Water, Biosphere II," but the real beginning of the project dates to 1998, when I spoke about these ideas with Doug Armato, then the new director of the University of Minnesota Press, and Carrie Mullen, acquisitions editor at the Press. Their enthusiastic response convinced me that it was time to make this project a priority, and I finally started working on it in earnest.

Further encouragement came from the four reviewers the Press asked to read and comment on my prospectus—professors Stephen Couch, Tim Luke, and Craig Humphrey, and environmental journalist Richard Hofrichter. Their comments were tough and challenging, and they forced me to think more carefully about certain core ideas so that I could articulate those ideas more convincingly.

As I was doing my first case studies, on the atomic fallout shelter panic of 1961 and on bottled water, I began to present these materials to groups of colleagues, at several meetings of the ASA, in various colloquium series at my campus, at the University of California at Santa Cruz (UCSC), at a colloquium at the University of California at Berkeley, and at several conferences held at the University of Wisconsin. I would like to thank every person who asked questions and offered critical comments during those presentations. Although it is never easy to hear criticism and I may not have been grateful for it at the time, I listened and learned. After those presentations, I would sit somewhere, madly scribbling notes to myself about what I should do differently or better.

Some colleagues offered specific comments; others helped by simply offering encouragement. I would like to thank, either for specific suggestions and criticisms or for general support, members of the Environment and Technology section of the ASA, including Stephen Couch, Craig Humphrey, Riley Dunlap, Harvey Molotch, Steve Kroll-Smith, Phil Brown, Bill Freudenburg, Timmons Roberts, and the late Fred Buttel. Similar thanks go to several colleagues in my department, Paul Lubeck, Helen Shapiro, Melanie Dupuis, Ben Crow, Wally Goldfrank, Craig Reinarman, and Herman Gray, and to other colleagues here at UCSC, especially Conn Hallinan, John Isbister, and Michael Brown. Thanks also go to Vern Baxter, good friend, former fellow grad student, now professor at the University of New Orleans, for his comments on the suburbanization chapter.

I would like to thank all the folks at the University of Minnesota Press who, at various stages, helped to move the project forward. That list would have to start with Doug Armato, the director, and Carrie Mullen, who was executive editor at the time the Press first agreed to publish this book. That list would continue with Jason Weidemann, my editor for the past two years, the person who has done more than anyone to bring this project through its final stages. The list would include Laura Westlund, Adam Brunner, Emily Hamilton, Catherine Clements, and certainly others who were involved but whom I do not know by name simply because I was not ever in direct e-mail contact with them. And the list would end with Mary Keirstead, who did such a fine job copyediting the manuscript. Thank you all.

Finally, my family. It took many, many hours, weeks that turned into months and months that turned into years, to research and write this book. Next to me, the highest costs were borne by my wife, Wendy Strimling, and my kids, Aaron, Emily, and Justin. As we all know, balancing work and family can feel like a zero-sum game. The deeper I let myself get into the flow of the work, the more distracted and unavailable I was to them. I am grateful that they understood, that they were patient and gave me the space and time I needed to fully immerse myself in this work. In addition, I thank Wendy for reading and commenting on one of the final drafts of the manuscript. She is an attorney, well trained in crafting airtight arguments, and she is also a skilled writer who is not shy about wielding the red pen. She is a tough reader, but when I listen to her comments, I always end up with a better product. I owe her double thanks, once for the sharp critique and again for the under-lying love and affection.

Introduction:
Inverted Quarantine

Not that long ago, hardly a generation back, people did not worry about the food they ate. They did not worry about the water they drank or the air they breathed. It never occurred to them that eating, drinking water, satisfying basic, mundane bodily needs, might be dangerous things to do. Parents thought it was good for their kids to go outside, get some sun.

That is all changed now. People see danger everywhere. Food, water, air, sun. We cannot do without them. Sadly, we now also fear them. We suspect that the water that flows from the tap is contaminated with chemicals that can make us ill. We have learned that conventionally grown fruits and vegetables have pesticide residues and that when we eat meat from conventionally raised animals, we are probably getting a dose of antibiotics and hormones, too. Contaminants can be colorless, tasteless, and odorless, invisible to the senses, and that fact increases the feeling of vulnerability.

According to the Environmental Protection Agency (EPA), indoor air is more toxic than outdoor air. That is because many household cleaning products and many contemporary home furnishings—carpets, drapes, the fabrics that cover sofas and easy chairs, furniture made of particle board—outgas toxic volatile organic chemicals. OK, we will go outside . . . only to inhale diesel exhaust, particulates suspended in the air, molecules of toxic chemicals wafting from factory smokestacks.

Even sunshine is now considered by many a hazard. Expose yourself to too much sun and your skin will age prematurely. You risk

1

getting skin cancer. The ozone layer has thinned, making exposure to sunlight even more dangerous. The incidence of melanoma, the deadliest form of skin cancer, is on the rise.

The response has been swift. Everywhere one looks, Americans are buying consumer products that promise to reduce their exposure to harmful substances.

In 1975, Americans were drinking, on average, one gallon of bottled water per person per year. By 2005, the latest year for which we have data, consumption had grown to twenty-six gallons per person per year,[1] over seven and a half billion gallons of bottled water. Bottled water used to account for only a tiny fraction of beverage consumption, inconsequential when compared to soft drinks, coffee and tea, beer, milk, and juice. Today, after enjoying years of "enviable, unending growth,"[2] bottled water has become the "superstar [of] the beverage industry."[3] In addition, nearly half of all households use some kind of water filter in the home.

A couple of decades back, organic foods had only a tiny share of the overall food market. Organically grown foods were sold, typically, in small "health food" stores. They were hard to find, even if you wanted them. Few people did. But now, after years of 20 percent annual growth, organic food is mainstream. There are not only organic fruits and vegetables but organic breads and cereals, organic meat, fish, and dairy, organic beer, organic snack food. One can find organic foods in large, attractive, upscale chain stores, such as Whole Foods, and also increasingly in mainstream supermarkets. Safeway and Wal-Mart both sell organic foods.

Those who can afford it buy "organic" or "natural" personal hygiene products, shampoo, soap, makeup; "nontoxic" home cleaning products; clothing made of natural fibers; furniture made of real wood; and rugs made of natural fiber. There is a new ritual in America (at least in middle-class America): applying 30 SPF sunscreen to our children's exposed skin every morning before they go to school, to summer camp, or to the beach.

A Resigned, Fatalistic Environmentalism

It struck me at some point that this was a strange, new, mutant form of environmentalism. There is awareness of hazard, a feeling of vulnerability, of being at risk. That feeling, however, does not lead to political action aimed at reducing the amounts or the variety of toxics present in the environment. It leads, instead, to individualized acts of self-protection, to just trying to keep those contaminants out of one's body. And that is not irrational if one feels that there is nothing to be done, that conditions will not change, cannot be changed. I think, therefore, that we can describe this as a resigned or fatalistic expression of environmental consciousness.

I also think something similar has happened in response to threatening social conditions. I detect it in the fact that millions of Americans have opted to live in gated communities or to move to exurbia, as far away from social problems as possible, away from the problems of the cities, from the manifestations and consequences of poverty, from deeply troubled race relations.

I detect it, too, in responses to the problems of the public school system. If one thinks it is broken and one cannot fix it, what does one do? Move to a neighborhood with good schools, if one can afford that. Start a charter school. Go private and support vouchers. Home school.

The Opposite of Social Movement

People respond to threat in different ways. One can engage actively with the issue: start or join an organization that campaigns for reforms that address the issue; support candidates who say they will vote for legislation that deals with the issue; e-mail congressmen and senators before key votes. One can organize, protest. That is the politically engaged, social movement/activism response. The kind of response to threat I am beginning to describe seems the very opposite of that.

An *Individualized* Response to Collective Threat

Social movements are collective in their goals and in their methods. They define problems as collective, and they say that only systemic change can fix them.[4] If food has pesticide residues, antibiotics, and hormones in it, it is because of the way most crops are grown and most farm animals are raised in the United States. If tap water has in it hundreds of chemicals at low concentrations, it is because chemicals from farms, industry, and millions of households have been disposed of in ways that allow waste chemicals to find their way into rivers, lakes, and groundwater. Systemic threats require systemic solutions, something substantial, like raising crops differently, disposing of wastes better, and so forth.

Social movements embody the notion that solutions are achieved only through collective means. Social movements exhort people to join with each other, to act together to (in Richard Flacks's apt and compelling phrase) Make History.[5]

In contrast, the kind of response to threat that I am interested in is individualistic in both goal and method. A person who, say, drinks bottled water or uses natural deodorant or buys only clothing made of natural fiber is not trying to change anything. All they are doing is trying to barricade themselves, individually, from toxic threat, trying to shield themselves from it. Act jointly with others? Try to change things? Make history? No, no. I'll deal with it individually. I'll just *shop* my way out of trouble.

A *Consumeristic* Response to Threat

To shield one's self from harm in this way inevitably requires the purchase of special items. The second obvious characteristic of this kind of response to threat, then, is that it is a consumeristic response. Faced with the same threat, another person might inform themselves more fully about the issue, join with like-minded folks, try to raise public awareness about the issue, try to get the political system to acknowledge it and deal with it. That is responding to trouble in the modality of *citizen in a democratic society.* A person who buys some products because those products promise to shield them from trouble is not at that moment a *political* actor. He or she is, instead, in the modality of *consumer,* responding to a felt need—in this case the need to be protected from harm—by buying certain goods that promise to satisfy that need.

"Inverted Quarantine"

As I began to understand that bottled water was just one example of a more general phenomenon, of a distinct type of social act, each instance of which has certain characteristics in common, I thought I needed to give it a *name.* After some reflection, I decided to call it *inverted quarantine.* I recognize that it is not a particularly felicitous expression, but the expression has the great benefit that it situates the new concept as similar to, but at the same time different from, something we know quite well, that is, *quarantine,* the public health measure that society has used for hundreds of years to contain the spread of infectious diseases.

The activity I was thinking about is similar to traditional quarantine in that it involves processes of separation and containment to keep healthy individuals away from disease agents. But here is the difference: In its classic form, quarantine is based on the assumption that the overall collective environment is basically healthy. Risk comes from a discrete source, such as a diseased individual. The community protects public health by isolating the diseased individual(s), thereby reducing the likelihood that others will be exposed and the infection will spread. What if we inverted the dyadic opposition—healthy overall conditions / diseased individuals—upon which the logic of traditional quarantine rests? The new dyadic opposition would be diseased conditions / healthy individuals. The whole environment is toxic, illness-inducing. The threat is not discrete, is not just *here* or *there*, not just these persons and not others, so it is not possible to separate off the threat, to contain it, to quarantine it. Danger is everywhere. How are healthy individuals to protect themselves? They can do so only by *isolating themselves* from their disease-inducing surroundings, by erecting some sort of barrier or enclosure and withdrawing behind it or inside it. Hence the term inverted quarantine.[6]

Inverted Quarantine as a Mass Phenomenon

Inverted quarantine has a history. This way of responding to threat was "discovered" or "invented" long before anyone started to worry about toxic hazards in food and water. It was, first, a way of dealing with *social* threat.

If the essence of inverted quarantine is the act of erecting a barrier between self and threat, one can trace the practice back very far, indeed, probably to the earliest fixed human settlements where walls were put up around the perimeter to control who entered and, if necessary, to repel attack, and to the rise of significant social inequalities, which required ways inside settlements to separate ruling elites from everyone else. We might think of that as the "prehistory" of inverted quarantine, but of course that is qualitatively different from the contemporary form of it, which became possible only in the modern era, when individualistic modes of action flowered and economics took the form of commodity production.

In the industrial cities of the nineteenth century, wealthy elites relied on inverted quarantine methods to put distance between themselves and masses of urban poor and working people. It is telling that at the time the poor, the homeless, the unkempt, the desperate, and the unruly were referred to as "the dangerous classes."[7]

People of means could control their interactions with those less well off than themselves by either moving to the "country," to one of the first exclusive suburbs built for the very rich,[8] or if they stayed in town, by retreating behind walls and gates, frequenting only wealthy milieus, traveling in private carriages, and employing guards to physically keep members of the dangerous classes away from them.

Buying one's way out of trouble, erecting barriers, separating and distancing one's self from threatening social conditions were expensive. At first, only the truly wealthy, a tiny minority, could afford to use such means to shield themselves from trouble. Others, further down the class ladder, might have wished to emulate them but could not afford do so. Today inverted quarantine has become a mass phenomenon. Millions—many millions—do it. Two distinct developmental trends, acting together, are responsible for this transformation from an elite practice to a mass phenomenon.

Downward Diffusion

Inverted quarantine could not become an option for many until either incomes rose or prices came down. In fact, both happened.

Economic development created a large and reasonably well-paid middle class of managers, professionals, and white-collar employees. For a time, even some blue-collar workers were paid well enough that they could make the inverted quarantine choice, at least some of the time.

At the same time, the price of some big-ticket inverted quarantine items fell. The suburban home is perhaps the best example. Once within the reach of only a small, privileged minority, a combination of causes, ranging from new construction methods to favorable federal home loan programs, lowered the cost of suburban home ownership to the point that millions could afford to buy one. Today at least half the population lives in a suburb.

Inverted Quarantine Applied to Toxic Environmental Threats

As people grew concerned about what polluted air, polluted water, and contaminated foods were doing to their bodies, the logic of inverted quarantine turned out to be readily transferable to dealing with environmental threats. Environmental protection is the new frontier for the practice of inverted quarantine.

Here, too, the combination of decent income and affordability fuels the growth of mass markets in inverted quarantine goods. True, some items cost much more than their conventionally made counterparts. An all-natural mattress can sell for thousands of dollars, way more than the ordinary futon or inner-spring mattress. Even the more affordable items, such as organic meats and vegetables, are enough more expensive than their conventional counterparts that consumers who have only modest incomes do not buy them. Still, they are not *that* much more costly. A substantial fraction of the public, numbering certainly in the tens of millions, at least, do have enough discretionary income that they can afford to spend more for organic rather than conventionally grown food, install water filters in the home, spend a bit more for "organic" personal hygiene products, and so forth. As a result, some of these products have become true mass-market items; others have not quite achieved that status but are gaining in popularity.

Why Should the Growth of Inverted Quarantine Concern Us?

People acting individualistically, as consumers? If that were all there was to it, the phenomenon would hardly be worth noticing, much less writing about at length. Individualism, as a mode of experiencing and as a mode of action, is at the core of our culture. Consumption is too. Consuming occupies much of our time, attention, enthusiasm—passion, even. It is hardly surprising, then, to see people apply the logic of consumption to situations (such as the one we are considering—having one's health threatened by toxic chemicals in the environment) that, from a naive point of view, seem to have nothing to do with acts of shopping. So why should the growth of inverted quarantine be of interest to us?

Any behavior that is widespread is likely to be of sociological interest. If only a few people act in a particular way, that is not likely to affect the course of things much, if at all. If, on the other hand, many millions do it, that behavior or activity can acquire real sociological or historical force. Millions living in gated communities, tens of millions living in exurbia, tens of millions drinking bottled water, eating organic food, buying "natural" goods—this is not just an interesting phenomenon; it is a phenomenon that is likely to have *consequences*.

Not that these consequences are intended. Time and again,

sociologists have shown that actions can have—often do have—
unintended consequences. People who engage in these kinds of
behaviors, move to gated communities, drink only filtered water,
and so forth, intend only to take care of themselves and their loved
ones. They do not mean to have some kind of larger impact on
the world. But their actions could have consequences they did not
foresee, did not intend, and do not necessarily want. Indeed, that
is what I wish to argue in this book. Environmental inverted quarantine is worth studying because it is likely to have serious consequences, in fact, consequences of historical significance.

I
Two Historical Case Studies

Two Historical
Case Studies

All indicators show that each year more and more Americans buy inverted quarantine products, such as bottled water, water filters, organic foods, and "organic," "natural," or "nontoxic" household products and personal hygiene products. Consumers obviously believe that these products will help shield them from toxic substances in our environment. As I wrote in the introduction, I am concerned about the long-term consequences for our society. What happens when many millions of people respond to environmental threats in this way?

Documenting the growing popularity of inverted quarantine products is easy. The evidence is overwhelming. Understanding the consequences is, at this point, a much greater challenge. The inverted quarantine approach to dealing with environmental threat is a relatively new phenomenon. The consequences are just beginning to be discernible. Since the direct evidence is scant, it would help to look at some other instances of inverted quarantine, instances that have a long enough history to have generated substantial consequences. I present two such case studies in Part I.

The first case I consider is the fallout shelter panic of 1961 (chapter 1). This episode had all the characteristics of an inverted quarantine event. Individualized response to collective threat? The scale of that threat was unprecedented, unimaginable. The end of society as we knew it. Half the population, possibly more, would be dead or seriously injured. All the great cities, the seats of government, commerce, and culture, would be destroyed; industry and agriculture,

11

devastated; the natural systems upon which all societies depend, poisoned. Yet, for a few months, quite a few Americans convinced themselves—and many more came close to convincing themselves—that if they only had a backyard or basement fallout shelter, they could make it through an atomic attack, then come out, rebuild, and get on with their lives.

This event seemed to me an excellent example, even a paradigmatic example, of inverted quarantine. I soon realized, though, that it was not exactly what I was looking for. The panic was too short-lived. It ran its course, lost steam, and was gone before it could have much impact on the course of the Cold War.

Still, read correctly, the episode can offer some important clues. During the panic, antiwar activists argued that if an atomic war did start, shelters would prove completely useless. They also said that if large enough numbers of shelters were built, that would actually increase the likelihood of war. These critiques were, in effect, warnings about unintended consequences.

I then looked for another case, an example of inverted quarantine that would offer more than just clues, a case that had actual, observable consequences. I needed an event that had a longer history, an event or process that had gone on long enough that the tendencies inherent in it had had time to ripen and fully express themselves. After some investigation, I came to believe that suburbanization could be that case.

More than half of all Americans live in suburbs. As I will show in chapter 2, suburbanization was driven, in great part, by an inverted quarantine–type wish to distance oneself—*insulate* oneself—from "urban problems." The inverted quarantine concept also fits certain recent developments at the cutting edge of suburbanization today, residential growth at the suburban fringe, often referred to as "exurban sprawl," and the proliferation of gated communities. The world is a dangerous place. When living in the suburb no longer provides enough of a buffer from social ills, one can move farther out, to the exurban fringes of metropolitan areas. One can retreat into a gated community.

Suburbanization has never been a demographically neutral process. Disproportionately it was whites and people with higher incomes who got to leave the city, leaving behind an urban population that had a higher proportion of minorities, a population considerably less well off, on average, than those who had moved away.

White flight (and although this was never acknowledged and never turned into a popular phrase, *class* flight) notwithstanding, substantial numbers of whites and substantial numbers of middle-class and affluent people did not leave. They stayed, living in the better neighborhoods, working, enjoying the cultural and recreational amenities, the museums, the theater, health clubs, fine dining, high-end shopping. What would it take to make affluent residents feel secure living, working, and playing in an environment that also had poor people, people without jobs, people without homes, drugs, gangs, crime? Inverted quarantine measures, of course. Residential neighborhoods bristle with electronic surveillance. Their streets are patrolled by private security police. Workplaces and recreational and shopping spaces are architecturally designed to separate the haves from the have-nots and to discourage the latter from ever coming face-to-face with the former.

Suburbanization has been going on a long time, more than a century. It has completely transformed the nation's social geography. It has increasingly separated people by race and by class. It has affected people's attitudes toward others. It has affected social and political attitudes. These consequences are observable *now*, and they are, indeed, significant. Suburbanization is the case study that has the most to teach about inverted quarantine's long-term impacts.

1. The Fallout Shelter
Panic of 1961

On July 25, 1961, in a nationally televised speech,[1] President John F. Kennedy told the American people that tensions between the United States and Russia had reached new heights, had perhaps reached some sort of breaking point. Kennedy had just returned from a summit meeting with Premier Nikita Khrushchev. The meeting had gone poorly. Khrushchev seemed intent on forcing the United States out of West Berlin. And, Kennedy said, West Berlin was "not an isolated problem." The Communist threat was "world-wide . . . Berlin . . . Southeast Asia . . . in our own hemisphere." The stakes were immense: "The [fate] of the entire free world." And much *more*: the very "freedom of human beings" was at risk. Well, when something so fundamental—freedom itself—is at issue, one should be willing to pay almost any price to prevail:

> We are clear about what must be done—and we intend to do it. . . . These actions will require a sacrifice on the part of many of our citizens. More will be required in the future. . . . these are burdens which must be borne if freedom is to be defended. Americans have willingly borne them before, and they will not flinch from the task now.

Sacrifice? What sacrifice? *What* burdens should Americans willingly bear? What did those vague words mean? Kennedy finally said it plainly: "We have [the] sobering responsibility to recognize the possibilities of nuclear war."

Kennedy was telling citizens they should be ready to be bombed. He tried, then, to also reassure them. He wanted to offer some hope that an atomic war would not necessarily mean total and utter catastrophe. He said: "In the event of attack, the lives of those families which are not hit in a nuclear blast and fire can still be saved, if they can be warned to take shelter, and if that shelter is available." He said he would ask Congress for $207 million for civil defense. With that money, the government would identify "existing structures . . . that could be used for fallout shelters in case of attack" and stock them "with food, water, first-aid kit and other minimum essentials for our survival." With that money, "air raid warning and fall-out detection systems" would also be improved.

Kennedy spoke almost exclusively about public mass shelters. He made only one, rather oblique reference to single-family backyard fallout shelters: "In the coming months I hope to let every citizen know what steps he can take without delay to protect his family in case of attack. I know that you will want to do no less."

Subsequent events showed that this was an incredibly effective speech, though perhaps not exactly in the way Kennedy intended. How do you live for more than a decade knowing that life as you know it could all end, without warning, in a matter of hours, if not minutes? Mostly you don't; you retreat into denial. That is how you stay sane. But the president had just announced that we were on the brink of full-scale atomic war. Suddenly terrified, the public latched on to Kennedy's promise that he would soon let everyone know what they could do to save themselves and their families. Before the speech, the Office of Civil Defense had been getting about four hundred inquiries a month; after the speech, more than six thousand per *day*.[2]

Media frenzy fed the panic. Television, newspapers, and mass circulation news magazines were filled with stories about nuclear war, civil defense, and fallout shelters. On television, both regular programs, such as *The Today Show* and *The Jack Paar Show,* and numerous "specials" urged people to consider building, as the title of one CBS special put it, "A Place to Hide." Shelters were the cover story in the September 15, 1961, issue of *Life* magazine. On the first page of the article, superimposed on a photo of a mushroom cloud, *Life* promised that "you could be among the 97% to survive if you follow advice on these pages."[3] The advice? Build a home shelter. Several different basement and backyard shelters are pictured in that issue of *Life,* some costing as little as $200. The illustrations of shelter living suggest that space might be tight, but the routines of white, middle-class suburban life could continue, pretty much unchanged. In one drawing, a child reads a book and Mom makes the bed while Dad opens the shelter door and scans the sky for incoming bombers. In another, Mom tucks Junior in for the night while Sis ties a ribbon in her hair and Dad lights up a cigarette.

The University of Michigan Survey Research Center reported that "almost two-thirds of the adult American population claim to have read such information [how to construct and stock a shelter] in some media."[4] Half the public said they were considering building a shelter.[5]

At social gatherings, people talked and talked about fallout shelters. Should we build one? Are you thinking of building one? If we do, should we keep it secret from the neighbors? If there is a war and we get in our shelter, what will we do if neighbors who hadn't built one beg us to let them in?

Demand for shelters boomed. Local contractors reinvented themselves and miraculously became, overnight, fallout shelter specialists. Sears planned to begin selling a prefab shelter kit that had been featured in the article in *Life*. Companies rushed to market shelter supplies.

The panic took a Hobbesian turn. A Jesuit theologian, Fr. L. C. McHugh, wrote an article that asked what Christ's teachings might have to say about "the pros and cons of gunning one's neighbor at the shelter door."[6] Father McHugh concluded that using violence to keep others out of the family shelter "when the bombs start falling" was consistent with "sound Christian morality." Sure enough, some guys boasted to reporters that they would use tear gas, guns, even "homemade shrapnel, grenades" to repel anyone who asked to share their family's shelter.[7] Officials in Nevada and in Riverside County, just east of Los Angeles, told their citizens to organize armed vigilante groups so that they could fight off the millions of atomic refugees who would be fleeing Los Angeles, overrunning their communities.[8]

Eventually, after raging for several months, the frenzy subsided. Kennedy and his advisors worried that they had inadvertently triggered a "national panic" that was getting out of hand and was becoming a "disaster."[9] They decided it was time to calm people's nerves. It helped that tensions over Berlin had eased. Enthusiasm for shelters was further dampened by news stories about shady operators who took unwary customers' money, dug a hole in the yard, maybe even built some flimsy structure, then disappeared. Given all that, people began to come around to the anti-shelter message, put forth by religious leaders, peace activists, hundreds of professors, and antinuclear scientists, that shelters were a folly that wouldn't save anyone.[10] By October, the Gallup Poll found "only one family in thirteen . . . [still] giving the matter serious thought."[11] The shelter building industry collapsed.

* * *

The fallout shelter is as close as I have come to finding a pure, paradigmatic example of inverted quarantine. Not just paradigmatic; all the essential features of inverted quarantine are, here, grotesquely exaggerated. (Would it be wrong to say it was clinically delusional?) I was, in fact, first drawn to using the fallout shelter panic episode as one of my case studies because I thought that its very grotesqueness would bring out, in the starkest possible manner, inverted quarantine's most essential features.

Later, as I was doing the case study, I discovered a second good reason for choosing it. As I noted in the introduction to Part I, antiwar activists argued that shelters would prove completely useless if war did break out. Worse, they argued, if lots of shelters were built, that would actually increase the likelihood of war. Put together, the two arguments predicted deeply ironic consequences. If Americans gave in to panic and built fallout shelters, believing that shelters were effective, they would actually be bringing on war, at which point they would learn, too late, that scrambling down into their little sheltered space could not possibly save them. These warnings begin to suggest, I think, a more general theory of inverted quarantine's unintended consequences.

I will discuss the antiwar activists' warnings toward the end of the chapter. I first recall the historical events that led up to and created the perfect conditions for a mass panic.

Creating the Conditions for Panic: Atomic Anxieties and the Failure of Civil Defense

A New Kind of Dread

The day after the bomb was dropped on Nagasaki, Japan moved to surrender. Everyone was elated. The war was over. We had won! Thousands of men—sons, brothers, husbands—would now not have to die in a protracted and bloody invasion of the Japanese mainland.

But, according to historian Paul Boyer, the celebratory mood was soon overshadowed by a strange new feeling of dread.[12] The atom bomb had been developed in total secrecy. Now, suddenly, the nation was told about this awesome new weapon. It was orders of magnitude more destructive than any weapon invented before. It could apparently destroy a whole city with a single blast. Dropped only twice, it made an implacable enemy beg for peace. Edward R. Murrow, that icon of American journalism, observed that the bomb left "the victors with . . . a sense of uncertainty and fear, with [the sense] that the future is obscure and that survival is not assured."[13]

Scientists had figured out how to release a new kind of force, elemental, awesome. Winston Churchill said man had unleashed a "secret of nature."[14] A radio journalist proclaimed that man had tapped "the basic power of the universe."[15] Were humans morally or psychologically fit to manage such a force? How could we possibly be mature enough, wise enough, sane enough to handle such power?

Commentators turned again and again to mythological analogies as they tried to articulate this concern. The myths they chose to recall—Frankenstein, Pandora—already answered the question. Exploding the bomb was like Pandora opening that box, unleashing all sorts of misfortune upon mankind. Or, like Doctor Frankenstein, we had created a monster that would turn on us, kill us all.[16] Of course we were not mature enough, wise enough. Our technological prowess had far outpaced our ability to think clearly or to make sound judgments.

There was a second, more concrete worry. If we could learn to make the bomb, others could too. Boyer quotes journalists, scientists, theologians, military men, all expressing the same awful thought: the bomb could be dropped on us.

In the next months, the public learned a lot more about the bomb. *Life* magazine had photographs of the mushroom clouds over both Hiroshima and Nagasaki, and before-and-after aerial photographs of Hiroshima.[17] Those photos were sensational, to be sure, but they were taken from far away; one can make out nothing of what had happened on the ground. Then John Hersey published his eyewitness account of Hiroshima in *The New Yorker.* Block after block, the city lay in ruin. Dead bodies everywhere. Many more injured, in agony, some so badly hurt they were barely recognizable as human. Not just injuries, but scenes straight out of a horror movie: "twenty men . . . their faces were wholly burnt, their eye sockets were hollow, the fluid from their melted eyes had run down their cheeks."[18] Hersey described a doctor who survived, trying to help the wounded:

bewildered by the numbers, staggered by so much raw flesh, Dr. Sasaki lost all sense of profession and stopped working as a skillful surgeon and a sympathetic man; he became an automaton, mechanically wiping, daubing, winding, wiping, daubing, wiping.[19]

Days after the bombing, seemingly unhurt people began to develop mysterious symptoms, "nausea, headache, diarrhea, malaise, and fever,"[20] that would later be understood as the symptoms of radioactive exposure. Many received fatal doses and died in agony.

Millions read Hersey's account. That issue of *The New Yorker* sold out; later, *Hiroshima* became a best-selling book. *Hiroshima* was read in installments on ABC radio.[21]

Some atomic scientists believed that the American government should renounce any further use of the bomb. The way to get the government to do that, they thought, was to tell the public what it would be like to be on the receiving end of an atomic attack. Harold Urey offered this account in an article in *Collier's,* one of the nation's most popular weekly magazines:

thousands die within a fraction of a second. In the immediate area, there is nothing left standing. There are no walls. They are vanished into dust and smoke. There are no wounded. There are not even bodies. At the center, a fire many times hotter than any fire we have known has pulverized buildings and human beings into nothingness.[22]

Single bombs flattening whole cities at a time was terrifying enough. But the popular imagination did not stop at that. There were rumors about invisible "atomic rays." "Rays" would make you sick. Your children would be born with defects.[23] It was said that if enough bombs went off, the whole planet could become unfit for life, not just for people, but for *all* plant and animal life.[24]

Visions of buildings and people being vaporized; survivors gravely ill, giving birth to mutants; perhaps an end to life on Earth. Boyer is surely right that "the atomic-bomb . . . was a psychic event of almost unprecedented proportions . . . [which triggered] powerful currents of anxiety and apprehension . . . through the culture."[25]

Descriptions such as Urey's and Hersey's could only intensify fear about what would happen if our enemies got the bomb. It was quite a shock, then, when on September 24, 1949, Washington announced that Russia had tested an atomic device. Within a few years, Russia would have operational weapons. In 1951, *Collier's* devoted a whole issue to imagining how World War III, "The War We Don't Want," would be fought.[26] We win, but atom bombs devastate major U.S. cities, New York, Washington, Boston, Philadelphia, Chicago, Detroit, Los Angeles, San Francisco. One lurid illustration shows Washington in ruins, burning, after being hit by an atomic bomb.

The Public's State of Mind Was a *Strategic* Matter

The country's leaders considered people's dread of the atomic bomb a serious problem. To cite just one example, in a speech to the New York State Civil Defense Commission, Nelson Rockefeller, then governor of New York, reproached the American people for being unwilling to "face this thing . . . without terror and without blacking out, . . . without shivering every time somebody mentions the possibility of a nuclear attack."[27]

Why was the public's state of mind such an important problem? Relations between the allies deteriorated rapidly once the war ended. American leaders anticipated problems maintaining the partition of Europe between Western and Russian spheres of influence. The Red Army was there, armed and ready, just over the border, while it would be too costly, economically or politically, for the United States to maintain, indefinitely, large enough ground forces in Europe. So the United States would hold Stalin at bay not with tanks and troops but with the threat of atomic retaliation.[28] As John Foster Dulles, secretary of state, told the Council on Foreign Relations, U.S. foreign policy rested, ultimately, on our "capacity to retaliate, instantly, by means and at places of our choosing."[29] Dulles did not need to explicitly say retaliate with *atomic* weapons; everyone knew what was meant by "retaliate, instantly."

The anticipated sequence of hostile moves looked like this: Russia threatens Western Europe; the United States prepares to send its bombers; since no sane leader would risk having his nation nuked, Russia backs off.

But when America's atomic monopoly ended, other sequences became possible: Russia invades Germany; United States retaliates; Russia drops Big Ones on New York and Washington; all-out atomic

war. Or Russia invades Germany; United States threatens to re-
taliate; Russia goes on red alert; widespread panic in the United
States; American leaders back down, conceding Germany to the
Communists.

Obviously, a foreign policy based on the threat that we would, in
the last analysis, resort to nuclear weapons would work *only* if the
enemy believed that the American people were willing to endure a
nuclear conflagration, if that is what it took to support their government's foreign policy initiatives. Paul Tompkins, of the Naval
Radiological Laboratory, described the required attitude when he
told a congressional committee that, yes, all-out atomic war would
be "catastrophic," but "if the chips ever go down . . . we should be
able to take it if we have to."[30] If American citizens were willing to
"take it," the nation's leaders could stand tall on the world's stage,
free to pursue an aggressive foreign policy. They could "negotiate
from strength," which meant, really, that they would not have to
negotiate much at all; instead, they could insist that other nations
bow to America's wishes. As Rockefeller put it:

> When the President sits at that table, he should be able to reflect
> the confidence of his citizens—free men who are not afraid . . . this
> will be a major strength for our negotiations: that the American
> people aren't afraid; they are willing to face this thing.[31]

If, on the other hand, Americans *were* afraid, if they weren't willing
to "take it," how credible was our threat to use the bomb? That is
why political leaders were so concerned that the average American
citizen was not prepared to "take it," did not seem willing to "face
this thing." If there ever was a confrontation in Europe, Rockefeller
worried that "the American people may very well say, 'Is Berlin
really worth that to us?'"[32] The enemy could then call our bluff,
could even turn the tables and subject *us* to "nuclear blackmail."

Just a Bigger Bomb

Rockefeller was right to worry. People were terrified. They were
not prepared to "face this thing." To calm their fears, some officials tried to persuade them that the bomb was just not that bad.
It was just a bigger bomb. In *How to Survive an Atomic Bomb,*
Richard Gerstell, formerly a senior radiological safety monitor for
Operation Crossroads, a series of tests at Bikini Atoll, wrote that,
yes, an atomic bomb's blast effect will kill you "if you're close to

the middle of an atomic explosion and in the open," but if you are farther away, the blast is not *that* destructive; it is "like a sudden, extremely powerful wind." And, yes, the bomb's initial flash is "as hot as the surface of the sun," but it will only "blind you for a few seconds or minutes if your eyes are open and you are close to it—just as any strong light like a searchlight can blind you for a time."[33] One would think that it would be difficult to protect one's self from heat as hot as the surface of the sun. Not so: *"this heat can easily be stopped. . . .* It will not go through even some very thin things. . . . Loose-fitting clothes make an air cushion around your whole body which protects you. The heat flash will often bounce off light-colored clothes. . . . Keep your shirt on with sleeves rolled down. . . . and be sure to wear a hat. The brim could save you from a terrible face burn."[34] Gerstell dismissed people's fears about "atomic rays." Yes, "it is true that this 'fall-out' stuff gives off rays . . . but it is not likely to hurt you. . . . you can easily protect yourself from it."[35] "Rays" would not give you cancer or make you infertile or cause you to give birth to mutants.

Surviving an atomic war, Gerstell implied, was pretty much like living through a moderately severe storm, such as a tornado or a hurricane. You get together some basic supplies, a first-aid kit, flashlight with extra batteries, radio, fire extinguisher, a pail with a cover, toilet paper, and so on. When attack is imminent, you cover windows with canvas or plywood, shut off pilot lights, fill small pots and bottles with drinking water.[36] Then, "if it happens, . . . get to your safe place, *lie down*"[37] (Figures 1a and 1b). When it is over, you pitch in, help fight fires, give first aid to those who need it "according to the rules in the Red Cross or Boy Scout Handbook."[38] You give your kids a hot bath (Figure 2), take a shower, wash things in your house to get rid of that radioactivity.[39]

Promise Them a Civil Defense to Believe In

Convincing people that an atomic attack was not that big a deal did not seem to work. If the credibility of U.S. foreign policy rested, ultimately, on "firm public morale,"[40] on "free men who are not afraid," on confidence that we can "take it if we have to," some other approach would have to be tried.

Officials thought the answer was to offer citizens a civil defense system they could believe in. They would stop trying to convince them that an atomic war was going to be no different that a World

War II–era air raid. They would tell them frankly that it would be horrible, damage on a scale an ordinary person could hardly imagine; *then* they would convince them that civil defense *could* see them through an attack and that they could then recover, afterwards.[41] Guy Oakes argues that the real purpose of civil defense in the 1950s was "emotion management." The goal was to replace "irrational terror of nuclear weapons" with "a healthy and measured fear."[42] Do emotion management right, you solve "the problem of national will":

> a robust and prudent fear would serve as a useful inducement in motivating the public to . . . participate in civil defense. Civil defense would, in turn, provide the domestic support essential to the policy of containment. If the American people . . . were convinced that they could survive . . . the problem of national will would be solved.[43]

Life magazine was quite explicit, a week after Kennedy's speech, when it wrote that

> in the minds of most defense planners, [civil defense] is part of positive deterrence. They hold that a nation with shelters can stand up firmly to threats of nuclear blackmail while a completely exposed country might find it difficult to risk nuclear attack in defense of some less-than-total threat of, say, Berlin size.[44]

It was a risky strategy. Officials from the Federal Civil Defense Administration (FCDA) had to first confirm people's worst fears, then convince them that the government was developing a civil defense program that would actually work.

Scaring People—the Easy Part

The FCDA pursued its educational scare campaign with great energy. The public was systematically, even relentlessly, exposed to dire facts and ghastly images. The FCDA published pamphlets and produced films and slide shows. It crisscrossed the nation with its "Alert America" exhibits, "made up of posters, blown-up photographs, movies, three-dimensional mock-ups, and dioramas . . . [which] depict[ed] as vividly as possible the effects of an atomic attack on the United States."[45] It went into the schools with pamphlets and short cartoons of Bert the Turtle, who showed kids how to Duck and Cover.[46] To illustrate one pamphlet, printed and distributed by the millions, the FCDA took a sequence of stills from an Army training film that showed a typical suburban house, that icon of the 1950s American Dream, being crushed, burned, and completely and utterly destroyed *in seconds* by a test shot in the Nevada desert (Figure 3).[47] The back cover of that pamphlet has a drawing of what appears to have been a person, now blown up, nothing left but hat and newspaper, with the words, "make no mistake . . . CIVILIANS *can* be bombed!" (Figure 4). As Spencer Weart says, the FCDA "spread images of nuclear disaster more efficiently than even the atomic scientists had done."[48]

Val Peterson, head of the FCDA, authored an article in an issue of *Collier's* titled "Panic: The Ultimate Weapon?" Written ostensibly to help the reader *not* panic if that fateful moment ever came, the article begins:

> You have just lived through the most terrifying experience of your life. An enemy A-bomb has burst 2,000 feet above Main Street. Everything around you that was familiar has vanished or changed. The heart of your community is a smoke-filled desolation rimmed by fires. Your own street is a clutter of rubble and collapsed buildings. Trapped in the ruins are the dead and the wounded—people you know, people close to you. Around you, other survivors are gathering, dazed, grief-stricken, frantic, bewildered. What will you do?[49]

FCDA officials tested air raid sirens regularly. Radio broadcasts were frequently interrupted by ominous tests of CONELRAD, the Emergency Broadcast System. Annually, from 1954 to 1958, the FCDA organized nationwide simulations—war games—called "Operation Alert," so communities could practice what to do in case of a real attack.[50]

A Civil Defense System Was Never Built

Scaring people was easy. The second part of the Civil Defense Administration's job, fostering calm, confident resolve by convincing citizens that there was a way to make it through the next war, proved much more difficult. Impossible, actually. Even if a civil defense system had been built, people may or may not have believed that it would protect them. But the point is moot. The fact is, *no civil defense system was ever actually constructed.* Two reasons why: First, the arms race kept changing the nature of the threat that a civil defense system needed to address. As soon as FCDA developed one scenario, it had to abandon it and rush to embrace another. Second, no one, not Congress, not President Eisenhower, not the military, was ever really willing to spend the billions of dollars it would have taken to actually build a substantial civil defense system.

The Arms Race

To us, today, those early, crude attempts to reassure the public (recall Gerstell, here) may seem like a joke, lies so transparent they hardly seemed worth telling. We should remember, though, that at that time the superpowers each only had a handful of warheads and only World War II–vintage planes to deliver them. The loss of a handful of the nation's biggest cities would have been a calamity, certainly, but the very fabric of society would not (yet) have been at risk. True, talk of wide-brimmed hats and loose-fitting, light-colored clothing *was* absurd, but thinking that a civil defense system could help was not (yet) a delusional flight from reality. But as the arms race heated up, the damage nukes could cause grew ever worse. Weapons became orders of magnitude more destructive, the problem of radioactivity was better understood, stockpiles grew, and delivery systems were revolutionized.

The hydrogen bomb was a thousand times more powerful than the atomic bomb. This orders-of-magnitude leap in destructive power was no secret. During the press conference at which President Eisenhower and Lewis L. Strauss, chairman of the Atomic Energy Commission (AEC), announced the first successful test, the following exchange between reporters and Strauss took place:

> Q: ". . . when the H-bomb goes off, how big is the area of destruction[?]"
> A: "Well, the nature of an H-bomb, Mr. Wilson, is that, in effect, it can be made to be as large as you wish, as large as the military requirement demands, that is to say, an H-bomb can be made as—large enough to take out a city. [A chorus of "What?"] To take out a city, to destroy a city."
> Q: "How big a city?"
> A: "Any city."
> Q: "Any city. New York?"
> A: "The metropolitan area, yes."[51]

The *New York Times* headline read, "H-BOMB CAN WIPE OUT ANY CITY, STRAUSS REPORTS AFTER TESTS."[52] Below the headline was a map of New York City with immense concentric circles showing the area that would be devastated by a single hydrogen bomb. The outer circle, showing the extent of incendiary action for one bomb dropped on Manhattan, reached north beyond Tarrytown, New York, west almost to Morristown, New Jersey, south to Plainfield and Perth Amboy, New Jersey, and east to half of Long Island. A similar map, of New York City with damage circles, was shown on national television news.[53] Elsewhere, similar maps showed what an H-bomb would do to other metropolitan areas (a map showing what one H-Bomb would do to San Francisco and the Bay Area is shown in Figure 5).

Big as those blast and heat circles were, fallout would greatly enlarge the area each bomb would impact. Officials scoffed, at first, at fears about "atomic rays."[54] By the mid-1950s, the fallout hazard could no longer be so easily dismissed. In March 1954, the wind shifted after a Pacific H-bomb test, and radioactive ash "snowed" down on the people of Rongelap, Rongerik, and Uterik islands, more than 100 miles from the test site. Many Rongelapese fell ill. Nausea; vomiting; diarrhea. They developed strange skin lesions. Their hair fell out. Blood counts were abnormal.[55] During the same test, a Japanese fishing boat, the *Lucky Dragon,* sailed though the fallout plume, many miles downwind of the test. By the time the ship returned to Japan, crewmen were gravely ill with radiation sickness.[56]

These events prompted a more honest public discussion of radiation and fallout hazards. A Senate committee held hearings on the civil defense implications of fallout. Both the Atomic Energy Commission and the National Academy of Sciences issued new reports about the fallout threat.[57] Everyone now understood that the size of the area that would be affected by a full-scale nuclear attack would be much, much larger than previously thought. As Ralph E. Lapp, an atomic physicist who wrote frequently about civil defense matters, explained:

> Up to this point [civil defense officials] had worked and thought mostly in terms of circles—the symmetric patterns of primary damage from superbombs. . . . Now superimposed upon the great circles of H-bomb blast and heat, there were zeppelin-shaped ellipses which stretched far beyond the circles of primary damage.[58]

The actual size, shape, and direction of those ellipses would depend on wind and weather conditions at the moment of attack. Even if one knew where the bombs were likely to fall, what areas would or would not be contaminated with radioactive fallout could not be predicted in advance. On the other hand, every fallout map I have seen showed large portions of the West and literally all of the United States east of the Mississippi blanketed with fallout.[59]

And stockpiles continued to grow and grow. In 1950, the United States had 369 bombs; the USSR, only 5. By 1959, the U.S. arsenal had grown to 15,468 warheads; the Soviets' to 1,060.[60] When Congress held hearings in 1959 on "The Biological and Environmental Effects of Nuclear War," witnesses were told to base their testimony on the assumption that the United States would be hit by 263 warheads, delivering a total of 1,446 megatons. As I will describe below, witnesses predicted devastating levels of damage, the end of modern American society, if not of human society, in any meaningful sense of that term. Terrifying as these predictions were, they were based on the assumption that the USSR would use only a fourth of its stockpile of warheads. What would keep them from sending more?

Finally, delivery systems were constantly being revolutionized. The Hiroshima and Nagasaki bombs were delivered by propeller-driven B-29s. Jet bombers, B-52s, B-58s, and their Russian counterparts, were faster, but there would still be many hours between first warning and impact. Then, in 1958, the Russians launched Sputnik. It is hard to believe, today, that a tiny ball, beeping faintly, caused such a shock. It was not the satellite; it was, rather, the launch vehicle. Sputnik signaled the start of the era when the ICBM would replace the jet plane and the span of time between warning and detonation would shrink from hours to minutes.

By the end of the 1950s, if war broke out, within minutes of first warning ICBMs would have delivered hundreds of megaton-range warheads. Every large city, probably every medium-sized city, would be gone. Industries would be in ruin. Tens of millions would be dead. Millions more would be fatally injured. For survivors, there would be an instant end to the familiar routines of everyday life, work, community, the web of personal relationships, plans for the future. Getting through the next few minutes, the next few hours, would immediately become the only issue. Only later would one think about finding one's wife, husband, children or wonder if the next morsel of food would be safe to eat, the next drink of water safe to drink. Even if one got through the day, a few days, sheer survival would be the only question for a long, long time.

Those were now what Gerstell liked to call the *facts*. As Dorothy said, in another context, "Toto, I have a feeling we're not in Kansas any more." The idea that one could, even on paper, design a reasonably workable civil defense grew increasingly unbelievable.

Civil Defense Scenarios Kept Changing

Civil defense officials struggled to keep up. As the threat evolved, they discarded whatever was their current plan and latched on to another. It hardly mattered; no one, not the president, not Congress, not the military, was ever willing to pay for *any* of the FCDA's schemes.[61]

The FCDA first advocated building *blast* shelters, structures strong enough to withstand the pressures generated by atomic explosions. Providing blast shelters for the entire population would have been prohibitively expensive, $300 billion by one estimate. In the early 1950s, that was a fantastic sum. Eisenhower, a real fiscal conservative, would never have supported spending such amounts, so FCDA officials pared down their proposal. The government should provide blast shelters only for several tens of millions, those who lived and worked in probable target cities, where the core of the nation's industrial economy was concentrated, and rather than build anything new, the government should only retrofit and strengthen existing structures in these areas. That radically reduced proposal was much less expensive, $250 million, not $300 billion. Congressmen still thought that was too expensive, especially since they were skeptical that blast shelters would do any good, anyway.[62] They declined to fund the FCDA's blast shelter proposal.

Soon the FCDA lost confidence in blast shelters too. Built sturdy enough, they could perhaps protect people from first-generation atomic weapons; they would be simply crushed by immensely more powerful H-bombs. But if there was enough warning, the FCDA's top administrators reasoned, you could get people out of the target cities before they were hit.[63] The FCDA began to promote evacuation as the main approach to civil defense.

But congressional hearings in 1955 produced evidence that evacuation would not work. Existing roads and highways could not handle the volume of traffic. Building bigger roads would cost billions. Even if those roads were built, jet-based delivery systems had reduced warning time to only a few hours. Mass evacuation would cause havoc. There would be gridlock, chaos, panic. Better understanding of the fallout hazard, following the contamination of Rongelap Atoll in 1954, was a further blow. As Thomas Kerr writes, "having only recently promulgated a policy that emphasized evacuation, civil defense officials were now faced with the unhappy prospect of evacuating people out of the prime target areas only to have them perish by radiation exposure."[64] The final coup for evacuation came just a couple of years later when the deployment of ICBMs reduced warning time to minutes.

Evacuation no longer a viable scenario, the FCDA veered back once again toward advocating shelters, this time *fallout* rather than blast shelters. At ground zero nothing could be done, but perhaps farther out, away from the big cities, far from blast and fire, fallout shelters could save lives. In 1956, the FCDA proposed a fallout shelter program that it estimated would cost $20 to $40 billion. Eisenhower balked at the price tag. He appointed an advisory committee, the Gaither Committee, to study the FCDA's proposal. The committee agreed with the FCDA; it lent its support for a $25 billion fallout shelter program. No matter; no one, not Eisenhower, not the chairs of the various congressional subcommittees that oversaw civil defense, not the military, supported spending such sums.[65]

Instead, the administration issued a National Shelter Policy, which said that the government would continue to provide information, pamphlets, "how to" manuals, and such, but that would be all. If an individual citizen wanted a shelter, he would have to build it: "Each property owner has an obligation to provide protection on his own premises. . . . Individuals are responsible for sheltering and sustaining themselves."[66]

In 1958, the Office of Civil Defense Mobilization (OCDM, the old FCDA, renamed) asked Congress for a modest $13 million to fund its responsibilities under Eisenhower's National Shelter Policy. Congress gave the OCDM $2.5 million, one *ten-thousandth* of the amount recommended by the Gaither Committee.

In a textbook on civil defense written for use at the Industrial College of the Armed Forces, Donald Mitchell offered this summary of civil defense at the end of the 1950s:

> after eight years of existence, no means had been developed by FCDA for protecting the population from atomic attack. . . . The blast and fallout shelters required to provide a reasonable degree of security in the face of nuclear attack are nonexistent.[67]

On the Eve of Panic

And so we arrive again on the eve of President Kennedy's July 1961 speech.[68] For ten years, officials had wanted the public to get scared, then prepare themselves for war. To their frustration, the public's response appeared to be one of "supreme boredom."[69] Scholars have sought to understand what people were thinking and feeling. They have looked at opinion poll data and other evidence and concluded that people's seeming apathy was a mask for complex, difficult feelings of profound terror combined with helplessness. Under normal circumstances, when war did not seem imminent, the terror and helplessness could be "ignored," and one could allow oneself to be absorbed by the routine concerns and challenges of everyday life.[70] But if tensions between the superpowers rose and leaders "rattled their sabers"—when the Cold War threatened to turn Hot—denial would suddenly cease working. People could no longer ignore the plain fact that they had *no defense* against the most terrifying weapons ever devised by man. Helplessness plus terror would combine in a different way and erupt as panic.

When Kennedy gave his speech, the situation facing the American people had two principal features: unimaginable destruction if war were to break out, and, in spite of all that talk, no civil defense system in place. Conditions were perfect for a panic, the social equivalent of huge quantities of combustible materials accumulating over the years in a forest. All that was missing was the burning match. Kennedy's speech on July 25, 1961, pretty much implying that we were on the brink of going to war, was that match, with results already described earlier in this chapter.

The Family Fallout Shelter
as Inverted Quarantine

In the family fallout shelter, all the essential features of inverted quarantine are present and in fact are present in the most extreme ways imaginable.

Collective Threat: An Average War

We need, first, to appreciate what an "average" atomic war, circa 1960, would have been like. Testimony from hearings held before the Special Subcommittee on Radiation of the Joint Committee on Atomic Energy in 1959 on the "Biological and Environmental

Effects of Nuclear War," plus a few other sources, together, provide an excellent overview of what was known at the time about the effects of nuclear war.

Seventy or so of America's biggest cities, all the major centers of population and industry, would be completely destroyed. About 50 million (in 1960 the United States had a population of 160 million) would die or be fatally injured the first day; another 20 million would be seriously injured. Burn experts testified to the "complete hopelessness of the problem confronting the medical profession." "Millions of severe burn casualties," only a fraction of all who were injured, "would overwhelm our capacity for adequate medical treatment." They said a "catastrophe of the magnitude that we're discussing today . . . is absolutely beyond comprehension."[71]

Gordon Dunning, chief of the Atomic Energy Commission's Radiation Effects of Weapons Branch, told the subcommittee that perhaps 5 to 10 percent of the surviving population would come down with leukemia and that all survivors' lives would be shortened by an average of about five years. Another witness, Robert R. Newell of the U.S. Naval Radiological Defense Laboratory, testified that fallout from the attack would cause "a lot of genetic mutations."[72]

Curiously, the subcommittee did not hear much about the nation's industrial sector. For that we turn to Herman Kahn's book *On Thermonuclear War* and Mitchell's war college textbook on civil defense. The subcommittee's scenario had bombs destroying seventy-one large cities. Kahn has a table listing what percentage of certain key industries were located in or near the fifty-three largest metropolitan areas of the United States: instruments and related products, 80 percent; transportation equipment, 77 percent; electrical machinery, 77 percent; primary metals industries, 77 percent; fabricated metal products, 72 percent; rubber products, 71 percent; machinery, except electrical, 66 percent; petroleum and coal products, 64 percent; chemicals and chemical products, 58 percent.[73] Energy is the key to any economic system. Concerning the American economy's two major sources of power, Mitchell wrote, "electric power . . . is apt to suffer severely. . . . The petroleum industry is exceedingly vulnerable."[74]

The subcommittee did hear quite a lot about agriculture and the food supply, primarily from R. F. Reitemeier, a soil scientist with the Atomic Energy Commission (AEC) and the U.S. Department of Agriculture. Reitemeier tried valiantly to put the best face on

a

b

Figure 1. (a) "Fall flat . . . cover your head." (b) "Crawl under a table or desk." New York State Civil Defense Commission, 195?: 16. Courtesy of Special Collections, University of California Library, Davis.

Figure 2. "Just wash off those radioactive particles." New York State Civil Defense Commission, 195?: 19. Courtesy of Special Collections, University of California Library, Davis.

Figure 3. "Atomic explosion destroys a suburban home." Federal Civil Defense Administration 1953: 2–5.

Figure 4. "Protect Your Family. Build a Home Shelter." Federal Civil Defense
Administration 1953 (back cover).

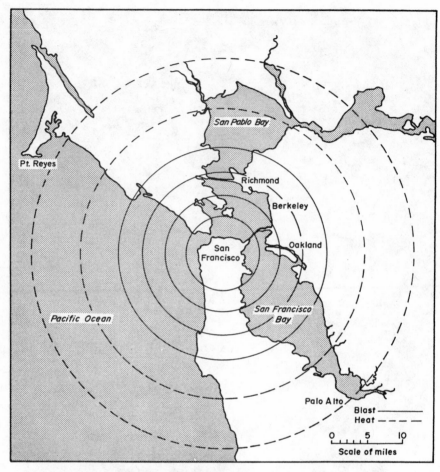

Figure 5. A 20-megaton H-bomb hits the Bay Area. Solid circles show blast effects; broken lines show heat effects. From Sidel 1966: 61.

Figure 6a. Inexpensive basement fallout shelter. Courtesy National Archives, photo no. 311-D-15-7.

Figure 6b. A $700 do-it-yourself model. *Life,* September 15, 1961: 104–5. Photograph: Dmitri Kessel/Time & Life Pictures/Getty Images.

Figure 7. Interior of an upscale underground home. From Swayze 1980: 33, 36.

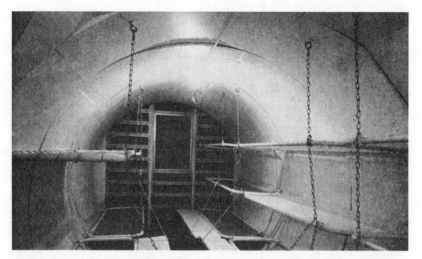

Figure 8a. Mock-up of one "room" in an urban shelter. From Wigner 1968: 40. Courtesy of Oak Ridge National Laboratory.

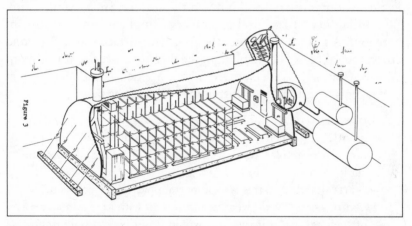

Figure 8b. Mass urban shelter design. U.S. Congress, Joint Committee on Atomic Energy, 1959: 689.

things, but his testimony showed there would be an immediate food crisis—survivors would either starve or they would be eating contaminated food—and long-term prospects for agricultural recovery were bleak.

Crops already harvested and food in warehouses and in markets could be washed, Reitemeier said, "if clean water were available."[75] (It is relevant to recall here Mitchell's statement that "a considerable proportion of all surface water probably would be contaminated."[76] Also see Guy Oakes's description of what it would take to have warm running water in homes after an atomic war, below.) Crops still in the ground would probably be contaminated too and in any case would be difficult to harvest due to the "unavailability of fuel and machinery."[77]

What about new planting? Fallout would have contaminated "virtually the entire region east of the Mississippi River . . . to moderate or severe degrees."[78] It was not hopeless, though. A farmer could plow deep, below the root zone, turn over the soil, then plant in the uncontaminated soil now on the surface. (One would think "unavailability of fuel and machinery" might be an impediment here too.) A farmer could "cut the sod," the surface soil and ground cover, remove it, and then plant in the dirt underneath. At this point Senator Clinton P. Anderson, from New Mexico, could no longer sit and listen as if all this was within the realm of the possible:

> SEN. ANDERSON: "May I stop you there and ask you what you mean[?] . . . When you say 'the sod,' are you referring to a pasture, for example? What would you do with the sod after you cut it and rolled it up?"
>
> DR. REITEMEIER: ". . . this is an important part of the problem."
>
> SEN. ANDERSON: "Well, what would you do with it? . . . in an area of such as . . . California . . . would it not be virtually impossible to cut the sod and remove it? Where would you put it? . . . You just could not possibly remove the ground cover from the Wisconsin pasture or Minnesota pasture and put it anywhere, could you?"[79]

If cutting and removing sod, roughly two tons of contaminated soil and ground cover per acre, was not practical, Reitemeier told the committee, one could plant, harvest, and discard the crop, repeatedly, until the radioactivity was gone. Unfortunately, that might take some time, "more than 40 crops . . . to achieve 90 percent decontamination of the soil."[80] Forty years of planting, harvesting, and discarding before one could harvest an edible crop.

Reitemeier's testimony showed, in effect, that any decontamination method one could imagine would require immense amounts of time, labor, and farm machinery just when those things would be in short supply. Characteristically understated, Chet Holifield summarized, "This certainly points up a tremendous problem for our food source lands in the event of a nuclear war, does it not?" Reitemeier had to concur, "It certainly does."[81]

The subcommittee heard from John Wolfe, chief of the Environmental Science Branch of the Atomic Energy Commission. Because Wolfe was not an antinuclear scientist/activist like Barry Commoner or, later, Carl Sagan, his analysis of ecological disruption is especially powerful. Wolfe's testimony was not about the ecological effects of radiation, which were still poorly understood at the time. It was not about nuclear winter. It was about nothing more exotic than forest fires, and all the more powerful because he focused on something so mundane, so commonplace; everyone knows about forest fires. The heat and blast from 1,446 megatons would, he said, set off fires that

> would spread over enormous areas of the dry western coniferous forests and in the grasslands. . . . [T]he dryer forest types of the Rockies and the Northwest and the Pacific Coast . . . would be pretty well burned. . . . In the Eastern United States, the dry oak and pine forests of the Blue Ridge and Appalachians from New England to Virginia, adjacent to multiple detonations, would undergo a like fate, as well as the pine on the southern Atlantic and Gulf Coast Plains.[82]

Wolfe then described some of the consequences in the mountains and the plains:

> With the coming of spring thaws . . . meltwater from the mountain glaciers and snowfields would erode the denuded slopes, flood the valleys, in time rendering them uninhabitable and unexploitable for decades or longer. Now, I do not mean to paint a spectacular picture here, but once the forces of erosion get into operation on three dimensional terrain, it is difficult if not impossible to check until there is some balance regained.
>
> Removal of the turf by fire and erosion on plains and prairie would result in uncheckable erosion by wind. I think it would be impossible

to estimate the area that would become dust bowl, but with the removal of the turf, the wind, which is omnipresent in the plains area, would form blowouts, and these would continue expansion.[83]

Wolfe's bottom-line conclusion was that

the effects are indescribable in their immediate implications, almost incalculable in their lingering results. . . . [I] believe that the survival of man or the rebuilding of his culture and his way of life will be impaired by environmental considerations and biological considerations which have been the result of nuclear attack.[84]

The hypothetical war analyzed by the Holifield Committee (entirely realistic, given the size of the superpowers' nuclear stockpiles) would have obliterated all cities of any significance. It would have destroyed most of the nation's industrial capacity. It would have contaminated much of the continental United States and have made it all but impossible to grow any crops. It would have profoundly disrupted the ecological foundations upon which, ultimately, every society depends. Survivors would emerge into a bleak world, indeed, "surfacing to death, chaos and despair."[85] They would find themselves "isolated on islands of survival, with devastated areas or areas of heavy fallout about them . . . cut off for an indefinite period."[86]

That is collective threat if anything is.

Individualized Response: The Delusional Belief in Radical Self-Reliance

If war means certain death, not only for individuals but for human civilization, the only really sane response is to try to stop it from happening—letters and phone calls to elected officials, petitions, mass demonstrations, getting officials to back off, de-escalate, negotiate. The shelter builder? Perhaps feeling powerless to do anything that could help avert the catastrophe, he acted alone (sometimes, even, in utter secrecy, so the neighbors would not find out). He made an enclosure and outfitted it. If the war came, he would climb in, lock the door, hope to get through the next two weeks.

Building a fallout shelter was individualism in a deeper sense too, because the collective threat was not just a couple of hours of massive destruction and two weeks of diminishing radioactivity; it was what Donald Mitchell referred to as "the problem of the 15th day," the aftermath.[87] It would not have just been a question of survivors finding loved ones, finding something safe to eat and safe water to drink, burying the dead, decontaminating, and then calling a community meeting so the survivors could pull themselves together and begin to rebuild. How does one rebuild a completely destroyed world, a world in which perhaps half the population is dead and the survivors in all likelihood injured, ill, in shock; not only no economy but no prospects of a reliable food supply for decades to come; no law and order;[88] ecological relations critically destabilized. Even Herman Kahn, the "strategic intellectual" who built a career on his willingness to think optimistically about the unthinkable,[89] conceded that thermonuclear "war is a horrible thing, and its horror lasts for some thousands, actually tens of thousands, of years."[90]

The impulse to build a shelter is not hard to understand. We all want to live. But building a family fallout shelter took more than a fleeting impulse. It took time and money, planning and effort. To do that, one would have to believe deep down, at least preconsciously, that it could work, that not only would the shelter get one through the attack but that one could come out and make a go of it even if society as we know it had been obliterated. The ability to believe that, even for a brief time, was possible only if one committed an act of profound denial. Several acts of denial, actually: To believe in such Crusoesque images of survival, one had to be in denial of all individuals' complete and total dependence on society, in other words in denial of the very *existence* of *society* as a complex set of interrelationships that are, as Émile Durkheim first taught us, *outside of* and *beyond* individuals but on whose continued existence and functioning every individual human life depends. One also had to be in denial of society's profound dependence on an ongoing, stable, sustaining relationship with nature. However wrong, however delusional—and that will be my topic in the next section— belief in the possibility of radical self-reliance, a faith in something like a deep, *ontological* individualism, was the necessary condition for making shelter building seem even marginally plausible as an answer to imminent atomic war.

I will concede that most shelter builders probably did not think too long or too deeply about what the "15th day" would really be like, but one can see that at some level such assumptions about self and society had to have been preconsciously taken for granted.

Those necessary underlying assumptions were articulated by some supporters of civil defense: Recovery *was* possible, they insisted, because Americans had it in them, individually and in small groups, to do what had to be done to get through it and rebuild! Eugene Wigner, the atomic physicist, talked of the American people's "inventiveness and ingenuity."[91] Herman Kahn talked of his "faith in the ability of people to improvise, to meet emergencies with some intelligence and energy." He wrote, "no matter how much destruction is done, if there are survivors, they will put *something* together."[92] After hearing the expert testify that most of the farmland east of the Mississippi would be contaminated, that there would be no fuel for farm equipment, and so on, Representative Craig Hosmer declared that "it will take much more than that to destroy my faith that the American farmers can produce surpluses under almost any conditions."[93] People would help one another, kind of like the way they come out to sandbag together when a river threatens to flood or the way they pitch in to clean up when their communities are ravaged by storm.

Commodification and Social Class

My initial definition of inverted quarantine says that it is a *commodified* response to risk. That might seem to be a point that requires no further discussion. To build a shelter, you had to first own a home, of course; renters and urban apartment dwellers had no basement or backyard where they could decide to put one in. It took money to hire contractors and builders. Even if a person did the labor himself or herself, building a shelter required building materials and tools. To outfit the shelter, one had to buy canned foods, medical supplies, and so forth. So, yes, having your own family shelter necessarily meant spending money, purchasing items. To that point, obvious.

The case study did bring to my attention, however, something that I had failed to see at first but now seems an obvious corollary to commodification—that anything that involves the purchase of commodities will necessarily have a class dimension to it. That was glaringly true of the family fallout shelter. Shelters were stratified along class lines, just like ordinary homes are.

In keeping with the Wizard of Oz imagery that surviving atomic attack was something like living through a big storm, the FCDA's early shelter drawings looked like the kind of structures that would get you through a tornado, and they could be built for a few bucks.

Dr. Willard Libby, Nobel laureate and once chair of the Atomic Energy Commission, boasted that he built a backyard shelter, "a hole in the backyard hill, bags of dirt, and some railroad ties," for only $30.[94] (Dr. Libby's shelter was destroyed by a brush fire that burned through his Bel-Air neighborhood shortly after he built it. This suggests that it may have not been sturdy enough to withstand a multimegaton detonation over Los Angeles.)

A homeowner who had only modest income could afford a small shelter either built into the corner of his basement or buried in the backyard (Figures 6a and 6b). *Life* magazine showed several such designs.[95] Typically, these cost less than $1,000.

Families who had more could build what was, essentially, a nice second suburban home underground. Jay Swayze, a developer from Plainview, Texas, designed underground homes/shelters intended for that class of shelter owner. A Swayze home would provide "safe harbor where a family would be protected in comfortable, familiar surroundings. . . . that equaled or exceeded present standards of living."[96] Swayze's homes had nice furniture, fireplaces, landscaped "yards," and "windows" in front of painted, backlit faux landscapes "a thousand times better" than the real trees and real sky most people see when they look out actual windows (see Figure 7).[97] A Swayze-built underground home was exhibited at the 1964 World's Fair.

Did very wealthy Americans also build shelters? If people in middle-class suburbs kept their shelters a secret, the very rich, who are typically far more secretive about their private lives, certainly would have kept mum about their shelters too. We know of only a couple of cases, via the stray anecdote. Swayze built a 16,500-square-foot underground beauty in—under—Las Vegas for Gerry Henderson, the founder of Avon. This home had everything—a generator, its own fuel supply, "an all-pink bedroom . . . crystal and gold fixtures in the bathrooms . . . swimming pool, a hot tub and hand-painted murals of outdoor scenery on the perimeter of the home."[98] It was on the market in 1996 for an asking price of $8 million. An article in the *New York Times* describes a now run-down and abandoned ruin of a shelter a *Times* reporter happened on one day while visiting the fifteen-acre Long Island estate of one R. Brinkley Smithers.[99] Arthur Waskow and Stanley Newman tell of a wealthy horse lover who in 1961 built a fallout shelter for her favorite stallion. They also mention that Nelson Rockefeller had a shelter built in his New York City townhouse on Fifth Avenue.[100] A single floor of that townhouse, a thirty-room triplex, was on the market in 1999 for $19.5 million.[101] I have not been able to find pictures of Rockefeller's shelter, but I very much doubt that it looked anything like the ones pictured in *Life*.

The class issue did come up once during the shelter panic, when JFK's inner circle reviewed the first draft of the "what you can do" pamphlet Kennedy had promised the American people. The draft's authors seemed to think that everyone in America was a white, middle-class homeowner. In a scathing memo to Kennedy, John Kenneth Galbraith wrote that the pamphlet

> is a design for saving Republicans and sacrificing Democrats. . . . [It is for] people who have individual houses with basements. . . . [not] for civilians who live in . . . tenements, or low cost apartments. . . . I am not at all attracted by a pamphlet which seeks to save the better elements of the population, but in the main writes off those who voted for you. I think it particularly injudicious, in fact it is absolutely incredible, to have a picture of a family with a cabin cruiser saving itself by going out to sea. Very few members of the UAW can go with them.[102]

The pamphlet was quickly revised, but no amount of official tact could deny the truth inadvertently disclosed by that first draft—that whenever response to risk requires the outlay of money, a person's class position inevitably determines both the extent and the quality of that response. If a man was deluded enough to believe in fallout shelters and had pots of money, he built a lavish underground home. Another homeowner, equally deluded but with less disposable income, made do with much more modest accommodations. Urban apartment dwellers had nothing. Spartan mass bunkers were designed but never built (Figures 8a and 8b).[103] Civil defense signs directed city dwellers to designated shelters in the basements of downtown buildings, some of which were provisioned with food and medicine (soon either looted or spoiled).

Not that it mattered in this instance. A rich person was probably just building himself a fancy tomb, a luxury crematorium. And if he did survive for two weeks, in his posh shelter, underground, he would then be forced to come out into a totally decimated world, and, like everyone else, would quickly perish.

Inverted Quarantine's Unintended Consequences

Can a study of civil defense and the fallout shelter panic tell us something about inverted quarantine's unintended consequences? It is not the perfect case for doing that because the panic ended before huge numbers of shelters were built, before shelter building blossomed into a mass practice. Nonetheless, I believe one can find some useful hypotheses in this history if one reads it with an eye for what might be inferred from it. Opponents of shelter building, a loose coalition of peace activists, antinuclear scientists, religious leaders, and university professors, developed two powerful arguments against fallout shelters. These were, in effect, predictions about the likely consequences of shelter building if the process continued to unfold and millions of Americans *did* go ahead and build them. I review those two arguments, then try to tease out the more general claims embedded in them.

Argument 1: Fallout Shelters Are a Cruel Illusion That Cannot Possibly Work

Critics argued that shelters would most likely fail immediately, during the actual attack. Warning time would be too short. Many people would never get home to their shelters. Even if they got there and retreated into their basement, surviving even the first few minutes would depend on where the bomb had hit. If their home was too near the point of detonation, the shelter and its occupants would be crushed, if not simply vaporized. A bit farther away, firestorms would either incinerate the occupants or suck air from the shelter, suffocating them. In shelters that remained intact, conditions would be awful, cramped, stuffy. Each person would be allotted only a few square feet of space. The food would barely be edible; drinking water, tightly rationed. After several days, sanitation would become a problem; body wastes would accumulate and could become a serious health hazard.[104] What if someone in the shelter fell ill?

The psychoanalyst Erich Fromm raised the question of "the state of mind of those in the shelter."[105] People would need to be in the best possible psychological condition if they were to cope with being confined together for two weeks while, outside, the world was falling apart. Instead, Fromm and his coauthor wrote, "Psychiatric experience and a huge body of literature" suggests that people in shelters would be experiencing

> traumatic neuroses . . . produced . . . by sudden fright and by tension of an intensity which transcends the amount our nervous system can tolerate. Such neuroses can result in severe depression, suicidal tendencies, self-accusations, amnesia and disorientation, and states of anxiety.[106]

But let's put all that aside and assume that somehow people would get through the in-shelter period. There remained "the problem of the 15th day," coming out of shelter only to find oneself in an almost completely ruined world. Critics of shelter policy argued that because modern society consists of a vast network of "elaborate, delicate interdependencies,"[107] reality would trump denial, and rugged, Crusoesque survivalism would be revealed to be what it always was, a cruel illusion. On the 15th day, as people came out and tried to cope with the simplest and most basic problems of survival, those interdependencies would immediately assert themselves in the *negative* form that the simultaneous malfunction of most, if not all, social processes would make it impossible to accomplish the most basic, the most mundane (but necessary) of day-to-day activities.

Take just one example. All the "how to survive" advice manuals recommend a hot bath or shower to wash away radioactive particles:

> take a bath or shower, if you have been in an area of lingering radioactivity. It is important that all radioactive materials be removed as soon as possible from your body, and bathing is the only practical means of getting rid of them. You won't need special cleaning compounds. Warm water and soap are ideal.[108]

In a brilliant passage in *The Imaginary War,* Guy Oakes begins to list some of the many things that would *still* have to be working right if survivors hoped to have hot water flowing out of their bathroom and kitchen taps following an attack:

> Plants would be in operating condition. . . . Highways would be open and streets would be cleared of debris so that workers could make their way to utility plants. . . . Workers on the way to their jobs would not be threatened by rioting mobs. Members of the labor force would be not only alive and uninjured, but also psychologically and morally prepared to appear at work at the appropriate time and

perform more or less as usual. Fear of further attacks would not keep the work force at home, perhaps boarding up the house, maintaining a furtive lookout for postattack looters.[109]

Such a list of necessary conditions could be extended indefinitely; each element mentioned by Oakes could be further "unpacked" and new ones added. Plants in operating condition? Not only would water treatment works, pumping stations, the whole network of pipes that distribute water to homes all have to be intact, there would have to be *power* to drive the water purification plants and the pumps. If water did get to homes, there would have to be gas or electricity to heat the water. And so on.

As Oakes says, considering all the things that would still have to be properly working "for water to flow when the American public turned on the faucet for the first postattack shower . . . the ability to take unlimited hot showers . . . was based on the supposition that a nuclear attack would leave American society fundamentally unchanged."[110] Essentially, *everything* would have to work like it used to. But nothing would be working. There would be no gas or electricity to run a water heater. There would be no utility companies pumping water into homes. There would be no hot baths or showers.

Similarly, agriculture today is completely dependent on ready supplies of pesticides, fertilizers, and irrigation water. Those inputs are present in sufficient quantities only if the petrochemical industry is producing and only if all the waterworks—dams, irrigation channels, pumps—are all intact and functioning. The petrochemical industry and waterworks, in turn, require a working energy system. Once crops are grown—if they can be grown—it takes an immense transportation network of railroads and trucks to move crops from farm to consumer. That infrastructure, too, has to be intact—workers at their jobs, fuel to run engines, and so forth. In essence, interdependence means that *practically everything* has to be in good working order today if one wants to be sure that there will be food on the table tomorrow. Thermonuclear war would simultaneously knock out most, if not all, of the subsystems that must function, all of them, together, if people are to have a ready supply of food.

I could continue in this vein, but the point should be clear. In a complex, modern society, each community, each individual depends, in every way, on products and services produced by a huge, complex economic system. If that system were largely destroyed—as it would be in an attack the size of the one analyzed by the Holifield

Committee—each surviving family and community would immediately face what it had not acknowledged by diving into shelter—
their profound dependence on the continued well-being of that
complex whole.

That was the gist of the antinuclear activists' first argument.
In their "Open Letter to John F. Kennedy," signed by hundreds,
university professors emphasized the need to distinguish between
individual survival and "the continued survival . . . of civilization as we know it, [which] requires protection of the physical
basis of society—the means of production and of distribution,
government, communications, etc."[111] In an influential speech
to the Commonwealth Club, Gerald Piel, publisher of *Scientific
American,* said civil defense is a "dangerous illusion" because it fo-
cuses only on individual survival and fails to acknowledge that individual survival depends on preservation of "the social and moral
fabric of western civilization."[112]

Herman Kahn said he was certain we could deal with all problems *individually,* the radiation, the loss of wealth, loss of half the
population, but what "if all these happened together[?]" Considering that, even he said "one cannot help but have some doubts."
Kahn conceded that his optimistic survival scenarios did not take
into account "interactions among the effects."[113] But that was
exactly the critics' point. All the impacts—destroyed cities; industry, agriculture, energy, transportation, and communication
mostly gone; millions dead and wounded; political system no longer functioning—would be simultaneous. In a catastrophe of this
scale, societal interdependence would express itself as "the *interaction* of the various disasters,"[114] making it impossible to deal
with any one disaster separate from everything else.

What about that good old human ingenuity? In an exchange
with Kahn, published in *Commentary,* Erich Fromm and Michael
Maccoby extended the interdependency argument to the depth of
the human psyche. People, they wrote, are dependent on society
not only for their material well-being but also for their very sanity.
How "would [people] react," they asked,

> to the break-up of the whole world around them[?] . . . survivors
> would witness a sudden tearing apart of the whole fabric of society.
> For most people, the sense of stability, and even their own identity,
> rests on the meaning society gives to their lives. . . . [Atomic war
> would destroy] a way of life which had given meaning to their ef-

forts, which had produced a sense of identity, as well as a sense of hope for the future.[115]

People would be dazed, overwhelmed, awash with grief and rage, just when mere survival required the highest possible level of functioning.[116] Privately, even the highest officials agreed. At a high-level meeting to assess one of the Operation Alert war simulations, President Eisenhower expressed the opinion that everyone, everyone from common citizen to the president, would be "completely bewildered . . . hysterical . . . going crazy . . . absolutely nuts."[117]

Mutual aid, pitching in, neighborly cooperation? David Singer wrote that the "post-attack social environment is as likely to be one of anarchy, plunder and murder, as one of order, restraint and cooperation."[118] Fromm and Maccoby thought so too, writing that social interactions would most likely be governed by the "ethics of the jungle and the concentration camp."[119]

Finally, looming behind all that was the prospect of ecological disruption. We have already cited the testimony of the AEC's own ecologist, John Wolfe, describing widespread fires, burnt forests, and water and wind erosion. Wolfe told the Holifield Committee that he believed "the survival of man or the rebuilding of his culture and his way of life will be impaired by [the] environmental . . . result of nuclear attack."[120] Speaking a few years later at a forum of the American Association for the Advancement of Science, Barry Commoner summarized the likely ecological impacts of all-out atomic war:

> epidemics of human and animal disease, crop destruction by radiation-induced ecological imbalance, erosion and sterilization of the land resulting from massive destruction of vegetation, and the triggering of possibly catastrophic climactic changes.

Commoner concluded that

> this nation, its population, its economic wealth, its social fabric— all that we speak of as civilization—would be irrevocably lost.[121]

Argument 2: Shelter Building Would Increase the Likelihood of War

Reviewing the debates that took place at the time about the wisdom of a national program to build fallout shelters, one finds a second critique: Critics of civil defense—the Federation of Atomic Scientists, the hundreds of university professors who signed the "Open Letter to John F. Kennedy," individual contributors to the *Bulletin of Atomic Scientists*—argued that a frenzy of backyard shelter construction would *increase* the risk of nuclear war. Shelters would foster a "false sense of security" that would make atomic war seem more acceptable.[122] Belief that shelters would protect people fostered "the delusion that thermonuclear war can be fought and survived," that "nuclear war might not be so disastrous after all."[123] Furthermore, the critics argued, the very process of shelter construction could "generate a sense of inevitability." That "fatalism," in turn, would "inexorably diminish the initiative and imaginativeness of our search for alternatives."[124]

Believing, falsely, that atomic war would not be so bad, the public would be more willing to have the government use nuclear weapons as an instrument of foreign policy.[125] Confident of public support, statesmen would be free to take a hard line in foreign policy, to take larger risks, to go to the brink.[126]

Furthermore, said the critics, if the Russians saw Americans by the millions building shelters, they could reasonably conclude that "we have markedly raised our own estimate of the probability of strategic nuclear war" and that we are preparing for a first strike.[127] That, in turn, could provoke a preemptive strike by the Russians.[128]

A false sense of security, a sense of inevitability, public acceptance of nuclear brinksmanship as foreign policy, American statesmen taking greater risks, paranoid Soviets—David Inglis concluded that "the net result seems to be that a substantial shelter program will increase a person's probability of being killed."[129]

At a symposium sponsored by the American Association for the Advancement of Science, some very prominent scientists agreed. The very logic of deterrence, they said, meant that a large civil defense program would inevitably accelerate the arms race and possibly trigger a first strike. Wolfgang Panofsky, later to become the director of the Stanford Linear Accelerator, explained,

> The principal usefulness of a "large" civil defense program in the strategic picture would be to enable the United States to consider a first nuclear strike against the Soviet Union as a response to threatening moves by the Soviets that did not involve a strike against the U.S. homeland. A "large" civil defense program would then limit damage in the United States to that which would be produced by a Soviet retaliatory strike. To be more specific, a large civil defense

program has the primary effect of releasing at least part of the U.S. population from its role as "hostages" in today's nuclear world.[130]

The Russians would want to maintain the credibility of their deterrence, to "defeat the purpose of civil defense" by building more, and more effective, weapons. Panofsky concludes that

> a large U.S. civil defense program might initiate another round of escalation that would lead to higher levels of strategic forces on all sides without an increase in our so-called "security."[131]

Owen Chamberlain, Nobel laureate in physics, agreed:

> So long as this policy of deterrence is used by both countries we must assume that each will make it a matter of major policy to maintain its deterrence threat against the other—the threat to do enormous and completely unacceptable damage to the other. . . . we must assume that meaningful attempts by the United States to protect its population from nuclear attack will be met by Soviet attempts to increase the effectiveness of their armaments. . . . The point of civil defense is an attempt to degrade the effectiveness of the adversary's weapons. It therefore forces a response on the part of the adversary. This means that the installation in the United States of a large civil defense system would cause the opening of a new round in the arms race.[132]

One might be tempted to dismiss these views as nothing more than what you would expect to hear from left-leaning social critics who opposed the use of nuclear weapons as a tool of foreign policy. But their arguments are only another way of saying exactly what Paul Nitze, Nelson Rockefeller, and other Cold Warriors said when they made a connection between public confidence and the pursuit of an aggressive foreign policy. If the American people were confident that they could "take it," the government would be free to pursue a hard-line foreign policy. The Right and Left differed only in their conclusions. The hawks thought widespread shelter building would build public confidence, hence geopolitical strength. The critics argued that if shelter building led to increased public acceptance of hardliner foreign policy, the result would be intensification of the arms race, greater suspicion between superpowers, and increased likelihood of war.

* * *

Well, none of that happened, happily. The panic subsided, and there *were* no catastrophic consequences, . . . making it possible for me to be here, writing about it, and for you to be reading about it, some forty years later. Everyone came to their senses. Most Americans decided, after all the nervous chatter, not to build a family fallout shelter. The leaders of both superpowers backed off from the brink of nuclear war and chose, instead, the saner path of test ban treaties and arms control agreements. Still, we can consider the shelter critics' arguments about what *would* have happened had millions of Americans decided to build shelters and ask what we can learn from them.

From the critics' first argument I take a more general point about the ultimate limits of individual self-protection. In the modern age, a radical individualism based on belief in the possibility of complete self-sufficiency is nothing but an illusion or delusion. Individualized self-barricading may work, to some degree, when problems are discrete and of modest intensity, when those problems do not rise to the level of threatening any of society's many complex, thoroughly interdependent subsystems, and hence all those subsystems continue to function reasonably well. If, however, those subsystems are stressed beyond a certain point and lose their ability to sustain the routines of everyday life, individuals quickly come face-to-face with the dependency they had up to then refused to recognize. In such extreme conditions, individual self-protection of any kind ceases to be viable.

The critics' second argument, that shelter building could itself have increased the likelihood of war? Although it is not simple or straightforward to generalize from the specific dynamics of international tensions or the psychology of two mutually suspicious, hostile nation-states on the brink of war, the fallout shelter episode presents us with one plausible scenario where mass practice of inverted quarantine could have helped drive society over a precipice. One may imagine other situations where the underlying process (the process generating the threat from which people are trying to protect themselves) is so pathological that not dealing with it, or just waiting too long to begin to deal with it, allows conditions to deteriorate to the point that a threshold is crossed and society spirals into profound crisis.

The relevance of such musings will become apparent in chapter 7. If we take seriously the scientific community's warnings that, as a global society, we are rushing toward an environmental crisis that in its own way could be as deadly as atomic war, then we may well be facing conditions where society's systems could begin to break down and all efforts at individual self-protection prove hollow. And if we conclude that there is a plausible causal connection between millions of people believing in inverted quarantine products and society's reluctance to acknowledge and address that crisis, then environmental inverted quarantine could be akin to millions believing in and building fallout shelters, where having faith in imaginary refuge helps bring on real catastrophe.

2. Suburbanization as Inverted Quarantine

At the beginning of the twentieth century, the United States was still overwhelmingly rural in character. Almost three out of four Americans—71.6 percent—lived on farms and in small towns. By 2000, that figure had dropped below 20 percent; 80.3 percent of Americans were living in metropolitan areas. Like other advanced industrial nations, the United States had become an urban society.

When one looks at that 80 percent more closely, though, one sees another development, arguably at least as important as urbanization: In 1910, less than 7 percent of Americans lived in suburbs; by 2000, that number had risen to 50 percent.[1] Central cities had grown tremendously, yes, but suburbs had grown faster. In absolute terms, the suburban population more than doubled between 1950 and 1970, then nearly doubled again from 1970 to 2000. Over the century, then, the United States had not just become an urban nation, it had become a suburban nation. In *Crabgrass Frontier,* one of the earliest and still one of the most often cited social scientific studies of suburbanization in the United States, Kenneth Jackson wrote,

> suburbia has become the quintessential physical achievement of the United States; it is perhaps more representative of its culture than big cars, tall buildings. . . . Suburbia symbolizes the fullest, most unadulterated embodiment of contemporary culture; it is a manifestation of such fundamental characteristics of American society as conspicuous consumption, a reliance upon the private automobile, upward mobility, the separation of the family into nuclear

units, the widening division between work and leisure, and a tendency toward racial and economic exclusiveness.[2]

* * *

Now, I am not an urban sociologist or a social geographer. I do not have an academic interest in suburbs per se. I decided to study suburbanization and its various sequelae because I believed it would help me understand what can happen when millions of people believe that inverted quarantine will protect them from harm.

As I will describe presently, suburbanization was to a great degree an inverted quarantine response to *social* threat, more specifically to distressing or threatening urban conditions. Americans have long been wary of the city. Jackson traces antiurban sentiment all the way back to the colonial era. He quotes Thomas Jefferson's opinion that "large cities [are] pestilential to the morals, the health, and the liberties of man."[3] In the nineteenth century, cities were widely viewed as places of "slums, epidemics, crime, anomie."[4] Andrew Jackson Downing, a tireless publicist for the suburban ideal, declared that "all sensible men gladly escape . . . from the turmoil of cities."[5] And he seems to have been right. When moving to the suburbs became affordable, all those "sensible men" did "gladly escape."

Later, as the first suburbs began to age, some of them, too, started to have what we think of as "urban" problems. Housing stock aged. Property values stagnated or fell. The suburbs started to feel a bit too crowded. Traffic got more congested. Crime rates crept upward. At some point, just living in the suburbs no longer felt like it was enough, so millions moved farther away, to exurbia, or withdrew behind the walls of gated communities. In other words, the feeling that one's initial inverted quarantine effort did not seem to be working led to further, more intense inverted quarantine activity.

Back in the city, affluent residents who stayed constructed protected spaces for themselves where they lived, worked, and played. That too can be understood as a new wave of inverted quarantine activity, made necessary by an earlier wave, because as the nation suburbanized (the first wave), cities were allowed to continue to deteriorate, requiring more intense efforts at self-protection by those who stayed.

I start with a brief review of the main features of suburbanization and explain in greater detail why the process should be understood as a series of increasingly intense inverted quarantine moves.

Toward the end of the chapter, I describe the suburbanization process's most important social and political impacts.

Suburbanization and the Transformation of Social Space

From Elite Beginnings to Mass Phenomenon

Until advances in mass transportation, such as commuter railroads and ferries, made daily trips into the city feasible, only the wealthiest families, those at the very top of the class hierarchy, could afford to deal with urban stresses by simply getting out of town.[6] A family had to have two houses, one in town and one in the country, used only some of the time, a place of respite to which one could occasionally escape. With the development of fast, affordable, regularly scheduled means for commuting, a family no longer needed two fabulous homes, one in town and one in the country, in order to have it all, both "rural" retreat *and* a job in the city. The family could live full-time in one of the new residential communities springing up along commuter railroad lines or near outlying ferry terminals, while dad rode to and from work every day.

One of the first of these upscale suburbs, cited as a paradigmatic example by every historian of suburbanization I have read, was Llewellyn Park, New Jersey, near Manhattan, developed in the 1850s. A thoughtfully designed planned community of big homes in a parklike setting that offered walks through woods, streams, and along a mountain ridge, Llewellyn Park was home to "successful businessmen and professionals who could afford an expensive residence and the time and cost of railroad commutation to Manhattan."[7]

Over the next one hundred years, a host of new developments, acting together, would drive the price of suburban housing ever lower and make it increasingly possible for middle-class, and eventually even working-class, families to realize this particular permutation of the American Dream. Innovations in home construction, such as the "balloon frame" house, and the use of mass production methods, such as those introduced by the Levitts when they developed Levittown, lowered the cost of building new homes. New modes of transportation, especially the streetcar and the automobile, had huge impact; each opened up vast new lands that were suddenly within practical commuting distance of downtown, making these areas attractive for subdivision and development. In addition, the farther away from downtown, the cheaper the land, so it was possible to offer a house for a lower price than the cost of a similar house in the city, or to offer more—more house, more yard—for the same price.

Federal subsidies, in the form of both favorable tax policies and home loan programs, contributed to making buying a home in the suburbs more affordable. The Federal Housing Administration's (FHA's) federally insured home loan program reduced the amount needed for the down payment, lowered mortgage interest rates, and stretched mortgage repayment to twenty-five or thirty years. With those changes, millions more could afford to buy a home. The Veterans Administration (VA) helped still more millions buy homes after World War II. Federal tax policy provided another powerful subsidy for home ownership by making both property taxes and mortgage interest deductible. Federal subsidies for highway construction not only gave a tremendous boost to the auto industry but also indirectly subsidized suburban residential development.

With all these subsidies, tax breaks, and home loan guarantees, Jackson writes, it "became cheaper to buy than to rent."[8] Purchasing a home outside the city was no longer a choice open only to people with lots of money. Suburbanization became a mass phenomenon.

Suburbanization of Jobs and Retail

At first, suburbanization generated a simple geographic division of labor—suburbs were primarily residential; the central city had the jobs and most of the shopping. Soon, though, both jobs (industrial work and white-collar office work) and retail began to move to the suburbs too.

Rail transport and, later, trucking lowered the cost of transporting goods, both raw materials to factories and finished goods to market, even over long distances. It thus became feasible to relocate industrial plants from city to suburb, and it was attractive to do so because land outside the city was cheaper. Industry began to move out of the city cores.[9] By the 1980s, two-thirds of manufacturing in the United States was located in suburban industrial parks.[10]

Corporate offices moved too, as suburbs began to offer nicely designed, newer office building complexes on landscaped grounds, tellingly called office "parks" and, eventually, "campuses." The suburban office park was an attractive alternative to offices downtown. Office workers and managers, many already living in the suburbs, appreciated the shorter commute. The fact that suburban white-collar employees no longer had to drive downtown every day also meant they no longer had to deal with downtown street scenes from which they felt increasingly estranged and which they found, at best, unpleasant and stressful and at times frankly scary. Cheap land meant office space cost less per square foot to build, own, or rent. A company could either save money or offer its employees a nicer, more spacious working environment. Everyone was a winner; no wonder corporate offices relocated in large numbers from downtown to suburb.[11]

Shopping moved too. At first, suburbs had only small retail outlets that offered only a limited variety of goods, the kinds of things a household needs every day. Later, small commercial nodes developed around commuter railroad stations and along streetcar corridors. Things really opened up, commercially, when suburbs started to be organized around the freedom of movement offered by the private automobile. Commercial strips lined major streets.

Then came the real revolution, the shopping center and the mall, with ample parking, shopping opportunities that ranged from specialty boutiques to big-name department stores, internal walkways where the temperature is always pleasant and it never rains, a food court. A person might still need to go downtown occasionally for some item or other, perhaps, but less and less often.

No longer just residential, the new suburbs had it all, homes, factories, office parks, multiplexes, every kind of shopping opportunity. They had become self-sufficient, had "liberated" themselves from central cities, so to speak. Some authors have argued that these developments have transformed at least some suburban areas to such a degree that they have, in fact, metamorphosed into some new kind of entity and it is no longer accurate to call them suburbs. A host of new terms have been proposed to capture the distinctiveness of this new species of social space: a patchwork space with residential neighborhood here, shopping there, an office park down the road, still elsewhere dining and recreation; nodes connected by roads and the automobile; not only no central city, no real center anywhere. Journalist Joel Garreau proposed the term "edge cities" for these new sorts of places. Mark Gottdiener and George Kephart suggest the more social scientific–sounding name "multinucleated metropolitan region." Robert Lang talks of "massively enlarged, growth-accelerated counties."[12] Others, sensing the unwieldiness of such phrases (and perhaps their unpleasant connotations of disease and cancerous growth) have come up with more sound bite–friendly expressions, such as "postsuburban regions" and "technoburbs."[13] Whatever the name, these suburbs-morphing-into-some-new-entity are, today, according to Rob Kling, Spencer Olin, and Mark Poster, "the most common form of metropolitan development in this country."[14]

Declining Quality of Life in Maturing Suburbs

As the twentieth century drew to a close, suburbs were still growing at an amazing pace. In the 1990s, the suburban sections of eighty of the biggest one hundred metropolitan areas grew faster than their central city or central cities.[15] Overall, these metro areas' suburbs grew twice as fast their central cities. Obviously, suburbs continued to be mighty attractive. At the same time, though, older suburbs, typically the ones nearest the city, therefore often referred to as "inner ring" suburbs, started to show their age, and they were no longer offering a way of life completely free of "urban" ills. They had grown "more diverse and crowded."[16] Their housing stock was aging and deteriorating. Property values were falling. Crime rates were rising. Even in the newer suburbs and in the sprawling "postsuburban" regions, continuing growth brought more traffic, more congestion, more stress in everyday life.

The desire to have the *real* suburban lifestyle inexorably drove people in one of two directions. Some moved farther out, to newer, presumably more purely suburban spaces, the kinds of new communities we think of when we hear the recently coined term *exurbia*. Others tried to recapture the desirable qualities of suburban living by retreating behind the walls and gates of so-called gated communities.

Outer Space: Exurban Sprawl at the Edge

Moving out to the edge of the metropolitan regions, or beyond the edge, to subdivisions newly carved out of desert, prairie, or forest, was one option, the option of choice for millions, judging by the demographic data. The fastest population growth rates in the nation are occurring at the exurban fringes as residential housing marches ever farther out, away not only from cities but also away from older suburbs. It is happening all over the nation, but most markedly in the South and the Southwest, in the regions demographer William Frey calls the "New Sunbelt."[17]

The neologism *exurbia* conveys the will to move so far to the margin that one is practically outside society (at least it physically appears to be so, but of course one stays connected via dish, cable, cell phone, broadband Internet access). Exurbia looks a lot like our stereotypical images of suburbia, seemingly endless vistas, subdivision after subdivision, of little else but single-family detached homes, dotted here and there with the occasional mall, big-box store, school, or church. Lang writes, "they are updated versions of bedroom suburbs."[18]

Inner Space: Gated Communities

If one does not want to move *so* far away, yet still wishes to have that sense of privacy, that *separation* from greater society (and its potentially uncontrollable impacts on one's life) that we associate with suburban living, one can live in a gated community. Gated communities are in many ways typical residential neighborhoods—well-kept homes, landscaped common space—only the whole of them is cut off from the surrounding neighborhoods by walls, and access to the inside is blocked by gated entryways. Residents open the gate electronically and enter; others must identify themselves before the gates go up or swing open and they are permitted to enter.

Early on, only the very wealthy could afford to live in a gated community.[19] Over time, however, homes in gated communities have become increasingly affordable, and they are now within reach for anyone with a decent, middle-class income.[20] Gating grew at phenomenal rates during the 1980s and 1990s, from roughly two thousand gated communities nationwide in the 1970s to about twenty thousand by 1997.[21] These twenty thousand communities have in them somewhere between three and four million households, about nine million individuals.[22]

Looking just at some Southern California data, one can see that exploding supply accurately reflects demand for such housing units. Supply and demand seem in extraordinarily good sync: The number of gated developments in Orange County doubled during the 1980s,[23] yet a 1990 survey of potential home buyers in Southern California showed that demand for homes in gated communities was far from satisfied. More than half of the respondents said they wanted a home in a gated community.[24] The market delivered: Edward Blakely and Mary Gail Snyder wrote in 1997 that close to 40 percent of new homes being built in California were homes inside gated communities.[25]

The greatest concentrations of gated residential communities are located in a band that runs from the Southwest east along the nation's southern flank, from Southern California, through Arizona and Texas, to Florida—essentially the Sunbelt. One finds them in large numbers in the suburbs of cities like Los Angeles, Phoenix, Dallas, Houston, San Antonio, New Orleans, Atlanta, and Miami, as well as in postsuburban counties that have sprung up beyond the suburbs of these cities. Gating is not just a southern or Sunbelt phenomenon, however; gated communities can be found in the suburbs of big cities in every part of the nation, in the New York metro area, outside Chicago, Washington, D.C., Boston, and Seattle.[26]

Is All Suburbanization Inverted Quarantine?

Suburbanization is a tremendously important phenomenon, clearly. It is critical to understanding many things about contemporary social life. Intrinsic importance does not, however, make it relevant or helpful to my inquiry unless suburbanization (and related phenomena, such as people moving to the exurban fringe or retreating to gated communities) are actually instances of inverted quarantine. Are they?

For basic suburbanization, the short answer is: At first only partly, then increasingly so as the process of suburbanization itself caused conditions in American cities to worsen, which then gave greater impetus for people to flee. For all the other developments I will discuss—exurbanization, gating, the construction of protected spaces for affluent urbanites—I believe it is completely appropriate to consider them manifestations of inverted quarantine.

The Beginnings of Suburbanization: The City Repels, the "Suburban Ideal" Attracts

I have consulted a number of histories of suburbanization, including works generally recognized as classics in the field,[27] and the consensus seems to be that the urge to move to the suburbs arose initially from a combination of "push" and "pull" motives. Urban conditions were appalling, filthy, noisy, crowded, stressful. The alternative—a place out in the "country," a quiet, peaceful neighborhood, people a lot like yourself, a pleasant, predictable, safe world—was very attractive.

Since people have been migrating to cities in huge numbers for hundreds of years, there is, obviously, plenty that is compelling about city life. There are jobs, all sorts of opportunities to make it economically. Cities are rich in other opportunities too: a greater variety of lifestyles, the possibility of new relationships, even the opportunity to reinvent oneself, to experiment with new identities. Cities have all sorts of attractions—vibrant streets, entertainments, shops full of incredible goods.

But cities also evoke dread and revulsion. I have already quoted Thomas Jefferson's low opinion of urban life. Kenneth Jackson argues that Americans have always been suspicious of the city.[28] Yes, in the popular imagination, cities were seen as exciting, filled with opportunity, but they were, at the same time, thought to be unhealthy, dangerous, immoral places where the innocent could easily slide into a dissolute existence.

This "anti-urban strain in American thought"[29] was based on something real. Conditions in the early industrial city were abominable. In the poorer districts, slum housing and overcrowding were the norm. There was as yet no infrastructure for collecting garbage and taking it to landfills outside of town, and no system of closed pipes to safely carry wastewaters away. Garbage—household trash, cooking wastes, worn-out clothing—was simply thrown into bins on street corners, where it piled up until it was scavenged or it rotted. Wastewater, water that had been used for bathing or to wash clothes, dishes, pots, and pans, was poured into open sewers out in the street. To this, add unregulated emissions from factory smokestacks pumping out columns of black smoke, columns that merged overhead to blacken the sky over the industrial city. The poet William Blake was not exaggerating much when he described early factories as "dark Satanic mills." At best, the city assaulted the senses. It was noisy. It smelled. Unsanitary conditions combined with other risk factors, such as poverty, poor nutrition, overcrowding, to create an ideal environment for disease.

Then there was crime, of course, and fear of crime. All the great cities were (are) characterized by astonishingly large differences in income and wealth. Some residents were fabulously wealthy; others abjectly poor. Far more of the latter, typically, than of the former. In the great cities of the nineteenth century "decent" folk were obsessed with the problem of the "dangerous classes,"[30] the homeless, the unemployed and the marginally employed, petty criminals, pickpockets, loiterers, beggars. Leading chaotic, desperately insecure lives, they had little opportunity or motivation to develop what we would now call internal controls or impulse control. They had little reason to be courteous or kind to anyone, especially to anyone better off than they. Present in large numbers, they helped create a threatening social environment where becoming a victim of crime was an ever-present possibility. One was constantly at risk, among strangers of questionable character. Who are these people? Will they leave me alone? Let me pass unmolested? Will they approach me, instead? Beg from me? Assault me? The propertied classes feared for their safety.

Many of the working poor were in poor health, the result of living in poverty, eating poorly, living in filthy apartments with insufficient ventilation, working too hard, too many hours inside hot, stuffy, airless factories. All these conditions combined were a perfect breeding ground for infectious diseases. So the rich also feared for their health.

And it was not just the wealthy, by any means. *Everyone* had reason to be afraid.

It is no mystery that people wanted to get away from all that. In addition, in the American case one has also to take into account the influx of immigrants from southern and eastern Europe and also the later migration of African Americans from the South. Racist attitudes toward these new urban residents, their different ways of life, their sheer presence, was a further reason for some white Americans to wish to distance themselves from city life.

But the impulse to go was not motivated solely by a wish to insulate oneself from unpleasant or threatening urban conditions. There were more positive motives at work too, *pulling* folks to the suburban way of life. In reaction to modernization and industrialization, a "rural ideal," already "deeply ingrained in the American national character,"[31] quite readily morphed into what the historians of suburbanization refer to as the *suburban ideal*.

Before industrialization, work was based in the household, embedded in domestic life. With the development of the modern economy, work was increasingly separated from the rest of everyday life. Separated from the rest of one's life, no longer under one's control, work was increasingly experienced as time spent in an "impersonal and sometimes hostile outside world."[32] One response to this change, endorsed and even championed by the churches, was to compensate by more intensely valorizing private life, family, home, and domesticity. Jackson describes how "ministers . . . in countless sermons . . . glorified the family [and] made extravagant claims about the virtues of domestic life."[33]

It was only a small step to the next idea, that in order to have family and domesticity one must have the proper physical setting, the single-family home. Widely read pamphlets and books redefined the home "as a retreat, a place of repose where the family could focus inward on itself."[34] The suburban ideal was born: owning your own home in a nice, quiet, safe neighborhood, a nice place to raise kids, a place that evoked nostalgic, highly idealized, airbrushed "memories" of an earlier, simpler time.

Other motives were present too, reinforcing the desire to move to the suburb. In a society that has some fluidity in class standing, status envy and the desire for upward mobility (and the equally powerful, if not more powerful, fear of downward mobility) predispose everyone to identify the "good life," the life one should yearn to have, with the lifestyles of peoples further up the class hierarchy. Recall that the wealthy were the first to opt for the suburban lifestyle. Inevitably, then, moving to the suburbs, to the ever nicer home, the ever more upscale neighborhood, became one of the great signifiers of success and social worth.[35]

At some point commercial boosters took over from the Christian ministers and the moral entrepreneur pamphleteers. Once you have built that commuter train or trolley line, you want paying customers. Once you have sunk a lot of your capital into buying land and putting up houses, you want home buyers.[36]

Over the years, the "American Dream" has meant many things—religious freedom, political freedom, personal freedom, economic opportunity. It is an ill-defined, therefore flexible, concept. It was not that difficult, it seems, to give it new content one more time, to sanctify the suburban ideal as the newest manifestation of the American Dream. So developers, real estate agents, railroad magnates, boosters of all sorts not only built homes, laid tracks, and so on, they also proclaimed that owning your own home—and if possible a series of ever nicer homes—in the suburbs was the new American Dream. A house in the suburbs was not just a home; it was *the* symbol of achievement of material success, status, *social* success.

One can say that this was classic boosterism, the intentional, calculated, manipulative creation of desire by the Captains of Consciousness;[37] still, one must admit that it went down pretty smoothly. After all, one can mount an advertising or public relations campaign and have no takers. People could have said no, no, we like the city and we want to stay. Prestigious social critics, such as Jane Jacobs, Lewis Mumford, and others, continued to sing the praises of city living and condemned suburbs as vacuous, dull, and bad for the soul. Movies, novels, and TV sitcoms made fun of suburbanites. It did not matter. Most Americans were sold on the idea, eager to move to the suburbs, if only they could afford it.

The Inverted Quarantine Motive Becomes More Prominent

Initially, then, suburbanization was driven by a complex set of motives, inverted quarantine–style impulses to buy oneself some distance—to insulate oneself—from threatening social conditions, but also by more positive, proactive rather than reactive, yearnings to have a life closer to what was envisioned as the cultural ideal. Later, though, as the process of suburbanization itself began to transform American society, the inverted quarantine component in that mix of motives grew in importance. The suburbs thrived; conditions in the cities worsened. The two developments were causally linked and must be understood not just as simply coinciding in time but as linked facets of a single process. And as urban conditions deteriorated, the inverted quarantine motive to leave the city became ever more prominent.

Thriving Suburbs, Deteriorating Cities

I have already described how over time the suburbs became more complex places, as industry, offices, and shopping all began to relocate there. Suburbs could "now perform . . . many social and economic functions which had [previously] existed as a bundle in the central city."[38] The suburbs' gains were the cities' losses. The rise of suburban manufacturing and retail commerce meant "the demise of the central city's economic monopoly in the 1950s and 60s."[39] The cities lost jobs. They lost some of their tax base.

And it was not just the private sector shifting their investments from cities, investing instead in the suburbs. The federal government was doing the same thing. Federal money that went for highway construction was money not spent on urban mass transit systems. Many miles of highways, whole grids of highways—an investment that indirectly subsidized suburban development—were built, while urban public transit systems were left to deteriorate.[40] Likewise, federal policies meant to help Americans buy homes, both the Federal Housing Administration (FHA) mortgage loan guarantee program and the Veterans Administration's mortgage assistance program (GI Bill), directed resources away from cities and toward building suburbs.[41] The FHA declined to insure home loans anywhere it judged that the "character of the neighborhood" might deteriorate in the foreseeable future. On paper, that seemed a rational policy. After all, home values in such neighborhoods were likely to fall, and it would simply be unwise, a bad business decision, to insure mortgages there. In practice, however, the decision rule was anything but objective and neutral. Change in "the character of the neighborhood" was a euphemism for more blacks and/or increased "occupancy [by a] lower class" of people.[42] These FHA decision rules (which were also used by the VA to decide where to assist home buying) strongly disposed the federal government toward insuring home mortgages in newer, suburban areas of a metropolitan area and disposed it against insuring mortgages in older city neighborhoods.

So, at the same time that the private sector was pouring money into building in the suburbs, the FHA and GI Bill mortgage policies were helping people buy those homes. In contrast, as Muller writes, "little public or private money was invested in the renovation of older homes within city limits,"[43] and that could only have one consequence, to "hasten [the] decay of inner-city neighborhoods."[44]

Outmigration "Sorted" Americans by Class and by Race

From the very beginning, the process of suburbanization exhibited powerful tendencies to organize social space along starkly segregated lines, first by class and nationality, later by race.

Income determined both one's ability to leave the city *and* the specific place in the suburbs where the outmigrating family would end up. Developers typically built whole subdivisions consisting of similarly priced houses. Since home mortgage payments are for most families the biggest item in the family budget, home prices served as the mechanism that sorted families by income, each gradation of family income with its own "patch," effectively segregated from others who had more or had less. As a result, the typical metropolis at the beginning of the twentieth century consisted of a core surrounded by a series of rings: an ethnic (i.e., immigrants from eastern and southern Europe) inner city; then a first, inner ring of older, "horse car" suburbs, already in transition from a middle-class to a working-class demographic; next the newer, "streetcar" suburbs, home to the middle class; and finally the outermost suburbs, with their mansions and exclusive country clubs, for the truly wealthy.[45]

Cities were disproportionately losing their affluent, their middle class, and even the better paid segment of their working-class residents. The poor, white and black, remained. True, many upper- and middle-class families stayed. On average, though, those who left had higher incomes than those who stayed. Over time, the disparity between the income of the typical suburbanite and the typical city dweller kept increasing. During the 1950s, during that first, feverish phase of postwar suburbanization, the difference between average suburban household income and average city household income rose quite substantially. The gap continued to grow, albeit at a somewhat slower rate, in the 1960s,[46] and the gap persists to this day. Census figures for the year 2000 show that in big metropolitan areas (population greater than one million), median family income in the suburbs is almost 50 percent greater than median family income in the central city. The differences are even more dramatic if one considers race. According to the census, non-Hispanic whites in the suburbs of the biggest metros earned about $63,000 annually, almost twice the roughly $32,000 earned, on average, by the inner-city blacks and Hispanics in those same metros.[47] In 2000, the poverty rate in the central cities of the 102 biggest metro areas stood at 18.4 percent; in the suburbs of the same metro areas, at 8.3 percent.[48]

The process of suburbanization "sorted" citizens not only by income but also by race and ethnicity. To some degree this was due to the strong correlation between race and class in America, but race and ethnicity have also operated as an "independent" factor by way of restrictive racial covenants (ruled unconstitutional by the U.S. Supreme Court only in 1948), de facto discrimination in the application of formally neutral and objective FHA loan guarantee policies (a practice commonly known as "red lining"), de facto discrimination embedded in the everyday practices of the banking and real estate industries, and white suburban homeowners' resistance to integration.

The combined effects of lack of economic means and racial discrimination ensured that few black Americans could participate in the great migration to the suburbs.[49] In their book *Separate Societies,* William Goldsmith and Edward Blakely sum up the results: "in the entire postwar period, suburbanization has been a racially biased affair . . . minorities have been left behind."[50] No wonder some pundits characterized the new social geography as "chocolate city with vanilla suburbs,"[51] and referred to the commuter trains and freeways that allowed suburban whites to bypass the inner city on their way to city jobs as "honky tubes."[52]

I will pause for a brief personal anecdote here. My family and I emigrated from Hungary to the United States in 1956. After a year and a half of hand-to-mouth living in New York City, my father found steady work in Kansas City. Soon after we arrived, the city's public swimming pools were desegregated. I recall watching it on TV. White Kansas Citians seemed pretty distressed to see black children splashing around with white children in those pools. A few years later, doing better, my parents bought a house in Mission, Kansas, a suburb of Kansas City. For six summers, until I left for college, I spent almost every summer afternoon at the Mission Pool, the public pool built for the residents of our suburb. In all those years, I *never* saw a black child at Mission Pool. Just now, thinking of my years in Mission, I took out my high school yearbook and looked at pictures of my graduating class, the class of 1965. I counted 695 seniors. Four were black. Two had Spanish surnames.

The differential nature of the outmigration could only contribute to making urban conditions even worse. The cities had lost manufacturing jobs and retail businesses, much of their economic base. Yet, even with resources shrinking, cities still had to provide services to society's most disadvantaged peoples, the poorly paid and the unemployed, minorities, and recent immigrants, whose numbers were not diminishing.

The Urban Underclass

The decline of the cities had the greatest negative impact on the lives of inner-city African Americans. The federal government was subsidizing housing and transportation in the suburbs, while neglecting housing and transportation in the cities. Housing stock deteriorated. Because schools are supported mostly by local property taxes, urban school districts' budgets fell; city schools deteriorated, while suburban schools thrived. Perhaps most important, the number of jobs available in the city plummeted. Blacks continued to come in great numbers from the South to cities in the Northeast and the upper Midwest. Lack of economic means combined with pervasive discrimination ensured that most of them would end up in the inner city, just when jobs were leaving the city.

The social consequences were devastating. In *The Truly Disadvantaged*, William Julius Wilson observes that "prior to 1960 . . . [black] inner-city communities [had] exhibited the features of social organization—including a sense of community, positive neighborhood identification, and explicit norms and sanctions against aberrant behavior," *in spite of* the fact that urban blacks were already suffering from high rates of unemployment and poverty and in spite of the fact that they had been viciously discriminated against for hundreds of years. Sometime around 1960, Wilson writes, something new began to happen: "Rates of inner-city joblessness, teenage pregnancies, out-of-wedlock births, female-headed families, welfare dependency, and serious crime" started to rise. By "the mid-1970s," Wilson says, rates of these very serious problems had reached "catastrophic proportions."[53]

Inner-city residents were trapped in abominable conditions that were "mutually reinforcing" and that set off "self-feeding spirals of decline."[54] Inner-city blacks were living in a world where "social disorder [was] inextricably woven into the fabric of daily life."[55] Although ultimately self-defeating, it is not hard to understand why they would respond with "modes of adaptation and [with] norms and patterns of behavior that [took] the form of a 'self-perpetuating pathology.'"[56]

After *The Truly Disadvantaged*, other books, with titles like *Separate Societies, American Apartheid,* and *Problem of the Century*,[57] continued to document conditions in the neighborhoods of the black urban underclass, the staggering levels of joblessness, the poverty and welfare dependency, the adverse effects on the institutions of marriage and the family, the high rates of violent crime.[58]

White Flight

Deteriorating conditions in the cities led to extensive rioting in the 1960s: Birmingham (Alabama) in 1963, Los Angeles in 1965, and Chicago and Cleveland in 1966; rioting in sixty cities, including Detroit, Newark, and Milwaukee, in 1967; and in Chicago again in 1968.[59]

A National Advisory Commission on Civil Disobedience (better known as the Kerner Commission) was hastily convened to investigate why the riots had occurred. The commission found that the riots were, in the last analysis, a response to conditions that had developed in the wake of a "division of our country into two societies; one, largely Negro and poor, located in the central cities; the other, predominantly white and affluent, located in the suburbs."[60]

The Kerner Commission's findings were hailed as an important warning that things must change. But the situation did not improve; it worsened.[61] To whites, the cities seemed to be becoming nothing more than "bombed out war zones."[62] This is the time the expression *white flight* was coined. During the decade of the 1970s, *every one* of thirty-six big U.S. industrial cities lost population because whites were moving away in such large numbers.[63] In the communities they left behind, segregation was reaching such new heights that to describe it adequately, Douglas Massey and Nancy Denton, analyzing data from the 1980 census, felt the need to invent a new expression: "hypersegregation."[64]

Positive Feedback and Inverted Quarantine

I think it is valid to think of suburbanization as having been one facet of a vicious cycle, or, more neutrally, a positive feedback loop: Initially people had a mix of different reasons for moving to the suburbs. They were "pulled" by visions of the amenities of suburban living and "pushed" by their wish to insulate themselves and their families from the stresses and dangers of city life. As suburbanization proceeded, urban conditions steadily worsened. As they did, the wish to get away inevitably began to play a bigger role in the decision to go. Frey captured this sense of vicious cycle or positive feedback loop when he wrote, "the central city effectively has been stripped of the metropolitan area's high income population and a good deal of its industrial tax base. . . . White flight . . . [is] largely a response to deteriorating economic and environmental conditions within central cities."[65]

Certainly, positive motives for wanting to live in the suburbs were still there, the wish to own a home, some grass and trees, less crowding, a safe place for kids to play. Those more positive motives—the "pull" of the suburb—did not disappear. The point I am making, though, is that though this outcome was not necessarily intended by the millions of individuals who moved out, suburbanization itself helped trigger a process that increased the force—the repelling force or "push"—to go because one had ever greater reason to want to insulate oneself and one's family from worsening conditions. Suburbanization helped create the conditions that would subsequently keep strengthening the inverted quarantine impulse.

It did not have to turn out like this. Looking at the "push" motivation (suburbanization motivated by flight from urban conditions), one can argue that there could have been another way: urban reform. Conditions could have been improved so that residents could enjoy all the vibrant benefits of city life without having to endure conditions that made city living unpleasant, stressful, or dangerous.

Consider, for a moment, some of the great cities of Europe. There, the inner city is the place to be, the place that has the nice homes, easy access to cultural amenities, shopping and dining, a lively street life. Residential suburbs at the edge of the European city are by comparison dreary, far less stimulating as environments, far less desirable; they are for the working class, for those who cannot afford to live in the center.

Urban reform was not just some hopelessly utopian dream. Real improvements were possible. Urban reform movements around the turn of the twentieth century, led by public health activists and other progressive reformers, managed to make significant improvements. Infrastructural investment created water treatment works that improved drinking water quality and piped that water into homes. Wastewater was piped away and treated. Garbage was collected and hauled to sanitary landfills.[66] Building codes improved housing stock. Urban parks were designed and built, greatly improving the quality of life of ordinary working folks who could not afford to travel to the countryside for relief from crowding, street noise, and foul air. Public transportation systems were built. Public libraries opened. Factory regulations and clean air and clean water laws all had positive impact on the urban environment, as did fuller employment, higher wages, and improved access to health care. A lot was done to make cities more humane, more livable, safer, healthier.

More could have been done too.[67] One can well imagine an alternative path of development in which policy continued to address other urban problems, thereby making the cities safer, cleaner, better places to work, live, and raise kids. That other path—the public provision of collective goods—was taken for some years, then largely abandoned as societal resources were diverted to building the suburbs. The cities were, essentially, left to flounder. Inevitably, urban conditions deteriorated, and the flight—inverted quarantine—component of the urge to move to a suburb, just as inevitably, grew ever more powerful.

Exurban Sprawl as Inverted Quarantine

Suburbs have begun to experience some of the same troubles that in earlier times were thought of as strictly "urban" problems: traffic, noise, rising crime rate.[68] As housing stock aged and became (relative to newer suburbs, farther away) more affordable, inner-ring suburbs grew less exclusive and more diverse. By 1990, the poverty rate in suburbs, nationally, stood at 8.0 percent (doubtless most of that concentrated in the older, "inner ring" suburbs); by 2000, it had risen to 8.3 percent.[69] In some large metropolitan areas, in states such as California and some places in the South, suburban poverty rates are well above 10 percent: in the suburbs of the Los Angeles metro, 14.6 percent; Bakersfield, 22.5 percent; Fresno, 19.6 percent; Miami metro, 16.0 percent; New Orleans, 13.1 percent; El Paso, Texas, 32 percent.[70] In states such as California, Texas, and Florida, in the region of the nation Frey has dubbed the "Melting Pot," Asian and Hispanic populations have been growing at such a fast pace that even the suburbs are now diverse, "almost as multi-ethnic as the cities."[71]

If a person does not like that kind of change, he or she can move to a suburb farther out, to a newly suburbanizing rural county, to exurbia. Now people move to the exurban fringe for different reasons, of course. In the early 2000s, in the midst of a housing price bubble growing out of control, economics was a big factor. Though not cheap, a home in California's Central Valley was much more affordable than a similar home in the Bay Area, so people bought homes there in spite of the horrendous commute. Still, the demographic data suggest that exurban sprawl should also be understood as a contemporary manifestation of white flight. Nationally, the pattern is clear. The farther out, the whiter the population.[72] Analyses of the 2000 census show also that many whites are leaving the Melting Pot region altogether, moving to the region Frey calls the "New Sunbelt," that is, certain parts of the West and the Southeast. When they get there, they move to the suburban fringe. In the New Sunbelt, Frey writes, "the fastest growth is occurring in outer suburban areas, exurban rural counties."[73]

When interviewed, exurbanites deny that their decision to move had anything to do with race. In an article in the *New York Times* headlined "Many Whites Leaving Suburbs for Rural Areas," one interviewee tells the reporter, "This wasn't like white flight or anything like that." The article then quotes Calvin Beale, a Department of Agriculture demographer: "They talk about getting away from urban crime, drugs, congestion and school problems. But it also means getting away from areas that have significant percentages of blacks, Hispanics and Asians."[74] For my purposes, it does not much matter if it is race, per se, or "urban problems," generically. The key is "getting away," the phrase that tips one off that we are in the Land of Inverted Quarantine.

Gated Communities as Inverted Quarantine

Moving to the exurban fringe is not the only option. Millions have chosen instead to live in walled and gated communities. Why do folks decide to live in a gated community? When asked, residents cite several reasons, but in the end, fear and the yearning for security always come to the fore as dominant themes.[75] Some residents say they want to live in a place that has a sense of community. Others talk about easy access to amenities within the walled community, such as tennis courts, a pool, a Jacuzzi, and so on. Still others say they want the status and prestige conferred by living in an exclusive neighborhood. Some residents say that a home in a gated community is more likely to keep its value. They are simply protecting their most important investment.

All perfectly reasonable and no reason to doubt that some or all of those motives are at work in the preference for living in gated communities. Nonetheless, the motive that comes through most insistently is that of security. The "outside world" is seen as a dangerous place. As they talk to social scientists interviewing them, the residents keep coming back to, dwelling on, stories they have heard about assaults, burglaries, rape, murder, missing children, property crime. Live out there in the open and you might find yourself in an unpredictable, unwanted, unpleasant, possibly even violent situation—even in a "regular" suburb.[76] The Dangerous Classes, it seems, are back. One can never be far enough away from them.

More broadly, there is a nonspecific, free-floating anxiety about the state of "things," a kind of global sense that the world is becoming ever more scary. It is that feeling, of vulnerability, of being at risk, that gives rise to the yearning for separation and for finding ways to *control* one's interactions with the social world. Walls and gated entries promise control, a life with no surprises.

All that is inverted quarantine, obviously. Collective conditions are hazardous and deteriorating? You can buy separation, control, protection. You can buy security.

Upon closer inspection even some of the other motives people offer for their decision to live in gated communities resolve into the inverted quarantine motive. Residents say they want a sense of community? The research shows that one does not really get close relationships in gated communities. One does not get the kind of rich, complex web of ongoing, face-to-face interactions that we imagine when we think "community." The typical resident in a gated community does not seem to interact with his or her neighbors any more frequently than the typical suburbanite.[77] The "community" that they find so comforting is, more exactly, demographic homogeneity, the reassurance and sense of security one gets from being surrounded by—not necessarily interacting with but simply living among—neighbors who are "similar to my background."[78]

If further proof is necessary that gating is inverted quarantine, one can go beyond the ethnographic data and consider some more "objective" correlates of the gating phenomenon. To wit: Gating, originally only for the elites, started becoming a mass phenomenon in the 1980s, about the same *time* that more traditional, open suburbs began to have some of the same problems only cities used to have, at a time, in other words, when simply living in the suburbs was no longer enough. Furthermore, gated communities are found in the greatest numbers in *places* where the population has become so diverse that even the suburbs have significant numbers of people of color, in the Melting Pot states of California, Florida, and Texas, and in some metros outside that region where suburban diversity is substantial, such as the New York metro (where suburbs are 32.6 percent nonwhite and Hispanic), Chicago (27.6 percent), and Washington, DC (40.2 percent).[79] Both timing and location support the inverted quarantine interpretation of gating because they suggest that people began to embrace living behind walls and gates exactly where and when merely living in a suburb was no longer enough to satisfy the wish to get away from, to insulate oneself from, unpleasant or threatening societal conditions.

Protecting Upscale Space in the City

The typical big American city today consists of two very different social worlds. The minority communities described earlier are still there, in the inner city, still un- or underemployed, mired in poverty, living in dire conditions. They have been joined, more recently, by millions of new immigrants, some of whom may someday follow the classic immigrant pattern and ascend the class ladder, but who, for now, swell the ranks of the urban poor. Nonwhite and Hispanic residents outnumber whites in the central cities of twenty of the twenty-five largest metropolitan areas, and in all ten of the largest metropolitan areas—Los Angeles, New York, Chicago, Philadelphia, Washington, DC, Detroit, Houston, Atlanta, Dallas, and Boston. In eight of those biggest ten metros, racial and ethnic "minorities" account for 65 percent or more of the central city's population. Poverty rates in the central cities of the ten largest metros range from 17.8 percent to 26.1 percent.[80]

The *other* urban world is the one inhabited by more affluent folks, who are also mostly white. Although whites have been leaving the city for decades, many stayed. They did not leave for exurbia. They did not retreat to a gated community. They are there, but they do not feel safe. Security is a fundamental issue for them. It is an issue in their residential neighborhoods. It is an issue where they work. It is an issue where they (and tourists and conventioneers) go to shop and play, in the downtown neighborhoods that have the upscale shops, the museums and art galleries, in the neighborhoods surrounding ballparks and basketball arenas, in and around refurbished historical downtowns and "festival marketplaces." Such places will attract visitors only if the people who would go there believe they will be safe and secure. These places are, however, often only blocks from streets teeming with the poor, the homeless, the mentally ill, the loiterers, from street scenes that seem to middle-class and upper-class folks rough, unruly, threatening.

Security—at least the *perception* of security—requires, first, the creation of clear boundaries between the two worlds and, second, procedures that will keep today's version of the Dangerous Classes out of the more privileged world. I looked for books and articles that would help me understand how these acts of partition and exclusion are, in fact, accomplished. I note, before I summarize what I learned, that the best articles I found were about Los Angeles, our second biggest city, where we find both worlds in extreme: Beverly Hills, Bel Air, Rodeo Drive, the massive Bunker Hill office complex, at one end; Watts, East L.A., vibrant immigrant communities, a city that is 70.3 percent nonwhite and 22.2 percent low income, at the other.[81]

In affluent residential neighborhoods, homes are fitted with sophisticated burglar alarm systems and bright outside lights activated by motion sensors. At some homes, the whole perimeter is fenced, and the driveway has an electronic gate with a surveillance camera. In many yards, a sign on the lawn warns that the home is protected by a private security firm. The sign threatens an "Armed Response" if there is trouble.[82] Private security is booming. In Los Angeles, the number of people employed by private security firms tripled during the 1980s.[83]

A residence is by definition private space. There is no problem, no ambiguity in drawing a boundary and declaring the inside off-limits. The other spaces that need to be secured, the office complexes and shopping and recreational places, present a greater challenge because they combine private, semiprivate, and public space in complex ways. Consider the inside of an office building or a shopping mall. Consider, further, the streets, walkways, and parking structures in an area of upscale shopping, or a tourist destination/festival marketplace, or some blocks downtown near museums. Controlling such space, defining boundaries, and either keeping undesirables out or, if that is not possible, keeping an eye on them and controlling their behavior requires, Mike Davis writes, both "an arsenal of security systems" and the "architectural privatization of the physical public sphere."[84]

If one has complete control over a development, as in the case of Bunker Hill in Los Angeles, one can simply "disconnect [the area] from the surrounding street grid,"[85] then architecturally design the whole complex, the buildings, the access, in ways that create a nearly impermeable border between the upscale, revitalized zone and the troubled, "poor immigrant neighborhoods that surround it on every side."[86] Steven Flusty describes how this border functions:

> The entire hill is . . . separated from the adjacent city by an obstacle course of open freeway trenches, a palisade of concrete parking garages, and a tangle of concrete bridges linking citidel [sic] to citi-

del high above the streets. Every path we try confronts us with the blank undersides of vehicular overpasses, towering walls studded with giant garage exhausts, and seating cleverly shaped like narrow sideways tubes so as to be entirely unusable. We could attain the summit from the south, but only by climbing a narrow, heavily patrolled stair "plaza," studded with video cameras and clearly marked as private property.[87]

In most instances, however, one cannot simply raze whole sections of a city and build the modern urban equivalents of castle walls, moats, drawbridges, and portcullises. Fortunately, more mundane security measures can be almost as effective.

First, elements of a complex must be connected. Trevor Boddy describes elevated walkways and systems of underground tunnels, in cities like Minneapolis and Atlanta, that allow one to go from parking garage to office, to stores, to lodging or entertainment without ever having to traverse the messy space of urban street life. "Bridges and tunnels allow the extension of the filtered corporate cities over entire sectors of downtown."[88] Atlanta's Peachtree Center, for example, connects a complex of hotels, shopping, and conference spaces with covered walkways high above the streets. Fly into Atlanta, take the airport shuttle to the conference complex. Once there, you never need to see sunlight or deal with any of a variety of social interactions that might arise in a city where 68.7 percent of the residents are not white and 24.4 percent are low income.

Elsewhere, elements are connected with "a succession of opulent piazzas, fountains, public art, exotic shrubbery, and comfortable street furniture along [the] pedestrian corridor . . . structures . . . intended to function as 'confidence-building' circulation systems that allow white-collar workers to walk from car to office, or from car to boutique, with minimum exposure to the public street."[89] As Boddy says,

> It [is] now possible to circulate through whole sectors of downtown on quarry tile and indoor-outdoor carpet, never encountering the sobering realities of concrete and asphalt; to walk from office to agency to restaurant under silkscreened banners waving in the pale wind of climate-controlled regularity; or approximate shopping-mall and home-and-school life in the heart of even the darkest downtown.[90]

Davis makes the same point:

> The new Downtown is designed to ensure a seamless continuum of middle-class work, consumption, and recreation, insulated from the city's "unsavory" streets.[91]

It is not enough, of course, to just create such privileged spaces. Undesirables must then be discouraged from entering or, better, even thinking of approaching. To some degree the border crossings are self-patrolled. People generally tend to not go where they are not welcome. As Davis writes,

> Ramparts and battlements, reflective glass and elevated pedways, are tropes in an architectural language warning off the underclass Other. Although architectural critics are usually blind to this militarized syntax, urban pariah groups—whether young black men, poor Latino immigrants, or elderly homeless white females—read the signs immediately.[92]

Social control cannot rely, though, solely on self-discipline, on have-nots voluntarily not going where they are not wanted. Borders bristle with security cameras. By one estimate, the number of surveillance cameras in New York City (65 percent nonwhite; 27 percent of households low income) tripled from 1999 to 2004.[93] Borders are patrolled by city police on the outside of the perimeter, and by private security on the inside. Here is how Boddy describes policing inside the Minneapolis skyway after social conditions worsened in the 1980s:

> the skyway system became . . . a fortress, a filter, a refuge. While the skyways had previously been policed only informally, security firms now received increased contracts to post officers whose purpose was to subtly dissuade the poor, infirm, black, native Indian, or mentally ill from entering the skyway system without explicit and closely monitored business.[94]

A border or boundary is always more effective if one can maintain a buffer zone extending out away from it. Officials can create a buffer zone between the actual border and the rest of the city—essentially create a hostile environment for street life or, to use the more pejorative term, loitering—by closing and locking city parks at night, eliminating places one can sit comfortably (for example, installing those cylindrical, half-barrel-shaped benches mentioned by Flusty), drenching certain spots during the night with sprinklers, and closing public toilets.[95]

None of this is exactly *new*. Whenever the wealthy have had to find a way to coexist with poor people in cities, they have used similar means to shield themselves from the unpleasantness or the danger of having to interact with those who are a lot worse off than they are. Walls, gates, bodyguards, private carriages, private schools, exclusive clubs—the specifics might differ but the function is the same. There is a qualitative difference, though, when the search for security is no longer just a matter of particular privileged persons' individual efforts but is instead "institutionalized in the very structure of urban space."[96]

Perhaps the thing to contemplate, though, is not that this is not new, but exactly that it is so old. After so many centuries, after so much talk of modernity's superiority over feudalism, of democracy and equality, of freedom on the march, of the transcendence of class (to say nothing of transcendence of caste), it should trouble us that observers feel that they have to resort to medieval imagery when they describe gated suburban communities, as in the book *Fortress America,* or urban redevelopment zones, as in Davis's article "Fortress Los Angeles."

Survivalism Lite

Off in northern Idaho, and elsewhere way off the beaten path, the Survivalists prepare for "end times." They have taken a long, hard look at the current situation and have concluded that all our problems are not just distressing but are *signs* of imminent collapse. They have assessed all the ways things could fall apart, from natural disaster to atomic war, to famine, to urban crisis and race riots. They have tried to figure out which parts of the country are least likely to be affected and have been retreating to those places for a couple of decades now.[97] They are preparing to make it on their own and to repel—violently, if necessary—the hordes of unprepared refugees who will be coming their way when social order crumbles and anarchy reigns, when we return to that pre-social, Hobbesian world of a war of all against all.

Articles in magazines like *American Survival Guide* discuss how to preserve and cache food, how to purify water, how to choose the right gun and knife, how to master combat tactics, how to "Get Off the Power Grid," and, generally, how to "Organiz[e] Your

Retreat."[98] Look at the ads. By mail order or via the Web, you can buy freeze-dried food in bulk (not in those paltry little packets back-packers buy at a camping store), water filters, generators, gas masks, military manuals, medical books, guns, knives, pepper spray, brass knuckles. As one headline has it, these folks are "armed, stocked, trained and ready for the failure of western civilization."[99]

From afar, the Survivalists look like extremists, if not lunatics. One might argue, though, that the difference between them and the rest of us is one only of degree. For Survivalists, inverted quarantine is *the* central organizing principle of their lives. Aren't they, though, only reflecting back to us in pure form what we are also, in fact, doing when we move to the exurban fringe, move to gated communities, or turn parts of cities into forts and bunkers? The only difference is that the Survivalists expect a definitive and total collapse, while the rest of us seem to expect that things will just kind of grind on as they are right now and all we have to do is to shield ourselves and our loved ones from its most obvious and im-mediate impacts.

Long-Term Consequences

Recall that I developed an interest in suburbanization because I thought I might be able to learn something about the long-term consequences of inverted quarantine. Having argued that subur-banization and other developments spawned by suburbanization are indeed instances of the inverted quarantine response to threat, I now try to identify the sociologically most salient impacts.[100]

Separate and Unequal Social Worlds

I certainly do not wish to idealize the past; Americans were seg-regated by class and by race long before suburbanization began. Nonetheless, I think it can be said that the process I have been de-scribing ensured that class and race segregation would continue to be the key organizing principle of social space as the United States modernized during the twentieth century.

Suburbanization dynamics ensured that suburbs and cities would have quite different levels of material wealth and resources. From the first, the average income of suburban households was higher than the average income of urban households, and the gap grew over time. In some cases, that gap, for example, between the most affluent suburbs and the inner cities, was very substantial indeed.

Then, the suburbs themselves are segregated, one from the next, by income. Homes in any one area tend to be similar in size and quality and are therefore similarly priced. Since mortgage payments are usually the biggest item in a family's budget (and there are cultural pressures to spend as much as one can to have the nicest possible home one can afford, in the best possible neighborhood), homogeneity in housing costs "sorts"—and thereby segregates—residents by income.[101]

Suburbanization sorted citizens by race and ethnicity as well. As Goldsmith and Blakely wrote, "suburbanization has been a racially biased affair." Huge numbers of urban white residents (though far from all) moved out, while "minorities [were] left behind."[102] Herbert Gans described the outcome, starkly, as "apartheid on a metropolitan scale."[103]

Meanwhile, back in the cities, better-off residents segregated themselves in their neighborhoods and also where they worked, shopped, and played, as I have previously described.

All these processes, together, have created a social geography that one can describe as "patchy": patches segregating people along both racial and social class lines, with great homogeneity *within* each patch, and significant inequalities *between* patches.[104]

Social Attitudes: The Perceptual Dichotomization of the Social World

What attitudes might flow from such an organization of social space?

The literature seems to suggest that suburbanization expresses above all a wish to turn inward, away from society. In his pioneering, still widely acclaimed critique of the suburb, Lewis Mumford famously described suburban living as "a collective effort to lead a private life."[105] Decades later, in *Crabgrass Frontier,* probably still the single most important social scientific study of suburbanization, Kenneth Jackson made essentially the same observation. The suburban way of life is a privatized life, with attention and emotional energies focused primarily on private matters inside the home with little left over for engaging with community. The same theme appears in recent works too. In *The Moral Order of a Suburb,* M. P. Baumgartner describes suburbanites' lives as by and large "encapsulated . . . within private homes and yards." Social ties between residents are "weak." There is little public life to speak of, and "street life is undeveloped."[106] Baumgartner observes that there are few overt conflicts between residents, which he attributes to a "culture of avoidance" and "the ease with which people can withdraw into their own enclaves."[107] Andres Duany, Elizabeth Plater-Zyberk, and Jeff Speck, leading theorists of what is known as the New Urbanism movement, argue that the very architectural design of the traditional suburb embodies, and reinforces, a withdrawn, individualized, privatized existence.[108]

These observations fit well with J. Eric Oliver's finding that suburban residents' political participation rates (as measured by such activities as talking to neighbors about issues and attending hearings) are strikingly low.[109] They also echo the themes of Robert Putnam's widely praised book *Bowling Alone,* in which Putnam traces a generalized decline in public participation, in "civic engagement and neighborliness," in modern America.[110]

Rob Kling, Spencer Olin, and Mark Poster have similar things to say about the quality of social life in postsuburban Orange County. Postsuburban "neighborhoods do not provide much support for casual 'hanging out' and thereby dilute the possible richness of public social life." Residents travel by car from residential neighborhood to office park to mall. Homes face away from the street; they open to the back, to private patios. "Such residential designs turn people inward toward the private spaces of their homes." All of Orange County seems to the authors to have been "implicitly designed to emphasize private domesticity and material consumption."[111]

Ethnographers of the gated community report that residents say they want "community," but such places deliver on the promise of community only if one defines that as living in a place where everyone is much the same in terms of income and, usually, race and ethnicity.[112] What gated communities do *not* deliver, they write, is community as a rich, complex, and vital web of daily face-to-face interactions. Neighbors do not spend much time together and do not feel particularly close to each other.[113] Evoking images of Russian nesting doll toys, Carol Tucker writes, "these communities promote privacy within privacy; residents tend to stay in their own backyards and do not visit on porches or front lawns."[114]

Although there is some, perhaps quite a bit, of truth to all that, this depiction of a nation of loners, individuals and families withdrawing into the privacy and isolation of their own homes, seems to me somewhat off the mark. I believe people *do* divide their world in two, into inside and outside, but the sociologically significant dividing line is not at the front door and the backyard fence. The line is not between self and world, between home and the rest of society outside the home. The line, I think, is actually farther away from self and home and encloses/includes a bigger social space that is familiar, safe. Inside that space, people work, socialize, shop. They belong to various church or civic organizations, homeowners' associations, the PTA. They may meet to discuss and make decisions about issues of immediate collective interest. Social spaces on the other side of the line, "outside," are experienced as stressful, tension-inducing, uncomfortable, scary spaces either to be avoided altogether or, if that is not possible, to be traversed quickly, with wariness, circumspection, and vigilance.

Indeed, if one theme pervades *all* the literature on suburbanization, exurbia, gated communities, and bunker living in cities, it is the omnipresent division of the social world into inside versus outside:

- For suburbanites, "inside" is the home community and other, similar suburbs; the city is the quintessential "outside." Back home, in their neighborhoods, suburbanites are "remarkably insulated from strangers."[115] It is familiar and safe. In contrast, suburbanites fear the city, which they see as dangerous.
- Discussions of the new "edge cities" or "postsuburbia" emphasize that living in one of these new hybrid places, with their own nodes of work, shopping, and recreation, means that residents never have to go to the city anymore.[116]
- Exurbanites are said to have fled not only the city but what they see as the increasingly citylike qualities of the older, inner-ring suburbs, with their traffic jams, rising crime rates, and too much diversity.[117] To them, even the suburbs feel like the "outside." Exurbanites are "a tribe of people who don't live in cities, or commute to cities or have any contact with urban life."[118]
- Residents of gated communities tell interviewers that they have chosen to live behind walls because the world outside those walls frightens them.
- In the city, public space withers or is abandoned to the forces of disorganization, while affluent residents keep to their controlled, forted spaces and stay well away from parts of town where there are large numbers of blacks or Latinos.[119]

- Some wealthy homes now have a "safe room," a secret, hidden room to which the owner can retreat if the home's boundary is breached, a further "inside" in the—God Forbid—event that the home is invaded and suddenly becomes part of the "outside." One can think of safe rooms in affluent homes as the urban fear analogue of the family fallout shelter.[120] Another instance of Russian nesting dolls.

This way of experiencing the world follows logically from the social geography of internally homogeneous, economically unequal residential patches. And it is perhaps only another way of saying what I have already argued, that the choice to live in the suburbs (and, later, in exurbia, in a gated community, or to retreat into a "gated and walled" way of life still in a city) is an *inverted quarantine* choice, expressing the belief that the world is a stressful, irritating, increasingly dangerous place and that the answer is to withdraw into more secure (also more pleasant, enjoyable) space, where things are predictable and under control.

Political Attitudes

The next question I wish to explore is, what *political* attitudes and behaviors follow from such *social* perceptions? I searched for studies of suburban residents' political attitudes and behaviors and found only a few,[121] and even fewer on the politics of folks living in exurbia and in gated communities. However, the studies I did find agree with each other to a great degree, and that increases my confidence in their findings.

According to these works, there is a quintessentially suburban form of political consciousness. Juliet Gainsborough suggests that it is best described as "defensive localism."[122] Defensive localism's positive side is concern for and a willingness to get involved with and materially support local issues. Its negative side is indifference or hostility toward problems or conditions that affect other communities or society as a whole.[123]

Defensive localism can be a matter of resisting redistributive policies. As Thomas and Mary Edsall write, suburbanites wish to hold on to what they have, to have the "highest possible return to themselves on their tax dollar."[124] That is great for any community that is affluent. It means better schools, nicer parks, a better library, a fine community center, athletic facilities. It also means that other, less well-off communities in the region, or elsewhere in the nation, will have fewer resources and will be less able to cope with their problems.

Defensive localism can take nonmonetary forms as well. Perhaps the most dramatic manifestation of the defensive side of suburban defensive localism was white suburbs' resistance to court-ordered busing. Plans to bus white suburban kids to inner-city schools and, conversely, to bus inner-city black children to suburban schools in order to achieve the desegregation goals of *Brown v. Board of Education* provoked tremendous opposition.[125]

Such political attitudes arise logically, perhaps inevitably, from the experiential split between inside and outside worlds and the patterns of associations and sympathies that arise from that. There seems to be enough sense of shared interests and enough contact—barely—*within* each "patch," especially within relative privileged patches, to sustain a politics of defensive localism. Beyond that limited world, though, the experience of everyday life I have described severely reduces opportunities to develop broader sympathies or understanding or broader, more abstract notions of self-interest that recognize how viable collective societal conditions are essential for one's ultimate personal well-being. People can, and do, spend almost their entire lives within their own demographically homogeneous patch. It is comfortable and safe inside; it is scary to venture *outside,* so why go there if you do not have to?

Understandable, yes, but such patterns of interaction have political/attitudinal consequences. As Frey remarks, "What is missing in this new scenario is the opportunity that used to exist for daily, face-to-face interactions among people from these different social worlds."[126] Voluntarily restricting themselves to being in their own communities, or to others that are similar to their own community, while making sure they stay physically, socially, *experientially* away from other, demographically different communities, people's sympathies and their understanding of their interests inevitably narrow. They lose a sense of how their own individual interests and collective interests are bound together. They lose (or fail to develop) a sense of responsibility for knowing about, caring about, and sacrificing time and/or money to maintain the shared institutions of social life. They lose (or fail to develop) the civic attitude that one is a citizen, a member of a polity.

Factor in too the fact that "public space," the places where traditionally people of different backgrounds met and interacted, has
either disappeared or has become so dangerous that people do not go
there. The degradation or the disappearance of public space is one
of the most common observations in this literature, whether the authors are writing about suburbs, "postsuburban" communities, gated
communities, exurbs, or the affluent, fort-ified parts of cities.[127]

If one hardly ever even *sees*, much less interacts with, other kinds
of folks, the most likely result is indifference or, worse, the reinforcing of negative stereotypes. Then "divergent interests and outlook
dominate perception. Inevitably, residents are mutually suspicious,
since they possess little understanding of each other."[128]

Imagine the political attitudes that sprout in such soil, in the kind
of social world we have been describing. In *The Moral Order of a
Suburb*, Baumgartner characterizes the general culture of the suburb as one of "moral minimalism": "weak social ties breed a general
indifference and coldness; . . . people . . . cannot be bothered . . . to
help those in need; . . . [There is a] lack of generosity; . . . a lack of
caring."[129]

If people hardly care about or feel generosity even for their *neighbors,* why would they care, then, or even think much about people
farther away, with whom they rarely, if ever, cross paths, whose lives
are nothing like theirs? Is it any surprise, then, that G. Scott Thomas
finds suburbanites "relatively indifferent to many of America's ills,"
and "not interested in funding new social programs"?[130]

Separating Politically as Well as Socially

In the early stages of suburbanization, cities continued to grow
outward by incorporating suburbs that had been built just outside
their current boundaries. Folks would move out of the city only
to then have the city annex the neighborhood to which they had
moved.

What good was it to leave the city if it could catch up with you,
gobble up your neighborhood? You may have physically removed
yourself to a nicer corner of the metropolis, but now you were
again part of the same political unit. You and your neighbors were
not in complete control of your destinies. Your taxes could go up if
officials needed more resources to address all the demands on the
city's budget. Those taxes would benefit someone else, in a more
distressed part of the metropolis. The city could force things on

you and your neighbors too: low-cost housing, or, worse, a half-way house, a waste treatment facility, or some other facility that the community needed but that you would not really want to live next to. Why retreat behind a symbolic moat if you cannot then raise the drawbridge?

Suburbs began to resist being annexed. They began to incorporate as separate and distinct governmental entities of their own. Incorporation, saying no to annexation, had important consequences. Before, when suburb and city were within the same political entity, different communities had to negotiate, compromise, work out how they would coexist. City government was the place where the different, sometimes opposing interests of the various peoples living in all the city's different neighborhoods were negotiated, where compromises were worked out. When suburbs drew political boundaries around themselves, when communities with very different demographic characteristics carved out their own little jurisdictions, their own city or county governments, the incentives to find workable ways to live together went away. As Oliver writes,

> Many suburban governments are constituted solely by people of one class, one race. . . . social conflicts that once existed among citizens are transformed into conflicts between local governments.[131]

Incorporating as a separate political entity, drawing a political boundary that corresponds to the social boundary between suburb and city, or between one suburb and another, facilitated the practice of "defensive localism." A community that is also a governmental entity can control its affairs, keep its own resources close to home, and keep "outside" society's problems at bay. The community can spend local taxes on itself and keep those taxes from flowing to other parts of the metro region.[132] For an affluent community, that means higher-quality local services, better parks, nicer pools, shorter lines at city hall.[133]

Incorporation makes it possible to give expression *in policy* to one's indifference to others' troubles. It makes it possible to refuse to help other communities cope with their problems. Kenneth Jackson writes,

> reject[ing] political absorption into the larger metropolis. . . . [having] the legal status of separate communities . . . peripheral neighborhoods [acquire] the capacity to zone out the poor, to refuse public housing, and to resist the integrative forces of the modern metropolis.[134]

A Tectonic Shift toward Conservatism at the National Level

Incorporation still leaves open the possibility that *federal* policy can transfer wealth from suburb to city. Defensive localism has to also express itself at the level of *national* politics.

Writing in 1968, the sociologist Herbert Gans observed that "state and Federal politicians from suburban areas often vote against anti-poverty efforts and other Federal funding activities that would relieve the city's financial troubles." Perceptively, Gans also predicted that the long-run consequence would be a vicious cycle, a downward spiral: "the economic gap between the urban have-nots and the suburban haves will only increase"; the cities, in ever greater need, would press for relief; that would then further intensify "suburban opposition to integration and to solving the city's problems."[135]

Writing about the same time but from the other end of the political spectrum, Kevin Phillips, a leading conservative Republican political strategist, opined that "suburbia and Great Society social programs [are] essentially incompatible. Suburbia did not take kindly to rent subsidies, school racial balance schemes, growing Negro immigration or rising welfare costs."[136]

Two contemporary book-length studies, by G. Scott Thomas and by Juliet Gainsborough, provide evidence that Gans's and Phillips's perceptions about where suburban political attitudes were heading were pretty much right on the mark. They find that suburbanites vote for candidates who condemn "big government" and "liberal tax and spend" policies. A closer look at the suburbanites' supposed antitax, antispend sentiments shows that they do not, in fact, oppose any and all federal spending. They support spending on schools, Social Security, and Medicare. Their opposition to taxing and spending is coded opposition to federal spending on cities, on the poor, and on blacks:

> Suburbanites say they oppose government waste, but they clearly do not oppose it across the board. Waste, in their lexicon, is defined as those programs that spend billions upon billions of dollars to help cities, minorities, and the poor. Equally expensive programs that primarily benefit the white middle class are not deemed wasteful; they are considered worthy of increased funding.[137]

Thomas contends that the "suburban middle class is . . . relatively indifferent to many of America's ills [and is] not interested in funding new social programs."[138] He points out that if one considers the New Deal and Lyndon Johnson's Great Society initiatives as expressions, at the level of formal policy, of a broadly based willingness to accept that a nation's citizens have a responsibility, collectively, for the welfare of minorities and the poor, the political views we have been attributing to a majority of suburbanites must be seen as a repudiation of those milestones in liberal social policy.[139] As Phillips wrote in 1969, "the burgeoning middle-class suburbs are the logical extension of the new popular conservatism of the South, the West."[140]

At the level of national politics, then, "defensive localism" has a natural affinity to hostile attitudes toward liberal social policy. That fact becomes ever more important for the overall tenor of national politics as the percentage of the electorate that lives in suburbs rises inexorably. Gainsborough calculates that the number of suburban congressional districts nearly doubled, from 88 to 160, in the twenty years between 1973 and 1993. In the same interval, the number of predominantly urban House districts fell from 78 to 67.[141]

Suburban political views push *national* politics toward the right, with major implications for social policies directed toward the poor, people of color, and urban conditions.[142] There is, obviously, an excellent fit between such attitudes and Republican Party political rhetoric. The suburbs vote Republican,[143] and recent Republican administrations, in turn, have been implementing the suburbs' political agenda. The Reagan administration's version of the traditional Republican battle cry for "fiscal conservatism" had nothing to do, obviously, with balanced budgets, for example; it meant instead cutting federal spending on social programs and cutting federal aid to cities.[144]

Gainsborough argues that in the 1990s the Democrats sought to compete for the suburban vote by moving rightward on social and economic issues, de-emphasizing their traditional support for redistributive policies such as aid to the urban poor, and embracing instead "welfare reform." That does not seem to have been enough, however, to win back the suburbs. In the 2004 presidential election, George W. Bush carried the suburbs by a margin of 52 to 48 percent.[145] The GOP's advantage in the burbs was only marginally greater than it was in the nation as a whole, but just look at the numbers from exurbia: according to demographer Robert Lang,

Bush carried 82 of the 101 fastest growing exurban "boomburg" counties.[146]

Although no single factor alone explains the recent rightward drift in American politics, I believe one can make the case that the changes in political attitudes that have come with the profound social-geographic transformation we call suburbanization have fueled and continue to fuel that trend. Suburbanization, then, not only generated long-term impacts *directly,* that is, driving the process of social-geographic transformation that gave us the United States we know today, a patchwork of homogeneous, segregated places, differing vastly in their levels of material wealth; it also fostered political attitudes that oppose and, in effect, veto any systematic attempt to politically address and ameliorate the conditions it helped cause.

* * *

As at the end of chapter 1, I can now ask what this case study teaches us about the possible consequences of mass enthusiasm for inverted quarantine solutions to collective problems. The consequences here are not as catastrophic as the consequences predicted by the shelter critics had Americans gone ahead and built millions of fallout shelters; they are, nonetheless, quite dismal, whether considered from the point of view of the health of society as a whole or from the point of view of the lives of individuals.

For the individual, residential inverted quarantine seems to work reasonably well, if one considered results only in a narrow and immediate sense. Federal crime statistics show, for example, that although suburbs are not completely free of crime, a person who lives in the suburbs is much less likely to become a victim of crime, especially of the more serious types of crime such as armed robbery or assault.[147] And for the most part, people do get many of the good things they hoped to get by moving to the suburbs: their own home, a bit of distance from the neighbors, a cleaner, quieter neighborhood, better schools.

In the bigger sense, though, creating and securing those small, safe worlds in the suburbs, in gated communities, in exurbia, and in secured enclaves inside cities come at a price. As inverted quarantine carves out a set of relatively protected places, societal resources are directed away from improving urban conditions, away from building more attractive cities, and away from providing everyone a job, a reasonable income, hope, self-respect. The result is social space organized as a patchwork of quarantined safe zones, on the one hand, and "no man's lands" that continue to be hazardous, on the other. On the one side are nice suburbs, upscale communities; on the other, deteriorated inner cities, Dangerous Classes.

The persistence of dangerous spaces and desperate peoples sets definite limits on how much safety inverted quarantine can provide. True, a person who lives in a gated community, carefully picks which roads to take when driving, works in a building guarded at the entrance, and plays and shops only in carefully chosen settings is a lot less likely to meet with foul play. Still, enclosure can never be complete. The unruly "outside" world can, on occasion, burst into the most guarded places. And there are times a person has to leave the safe zone and cross through dangerous spaces.

And there are psychic costs; one might even call them *existential* costs.

Since danger still lurks, you cannot fully enjoy even the partial protection you do achieve. There is no end in sight. Inverted quarantine measures are not one-time things; you have to keep at it indefinitely, on and on.

The more a person turns to inverted quarantine to deal with risk, the more it distorts and degrades that person's life. When inverted quarantine becomes one of the central organizing principles of a person's life, freedom of movement incrementally decreases. The person voluntarily imprisons herself or himself, restricting movement in social space. In such cases, this person achieves at most a partial benefit at the cost of literal quarantine in an ever shrinking life-world.[148]

In my social theory class, I have students read Michel Foucault's *Discipline and Punish*. Foucault argues that modern Western society is in some fundamental sense "carceral." I have to admit that I have always considered his claim somewhat over the top, Foucault indulging in a bit of theoretical hyperbole.[149] However, if one thinks of it not as a description of past and present but instead as naming certain tendencies immanent in our situation, the notion does not seem so extreme. "Carceral" might be an appropriate way to describe the logical end point of social inverted quarantine, and thus its unfortunate existential consequences. Gated communities, locked and surveilled buildings on one end of the social hierarchy; prisons, neighborhoods of the poor patrolled by police, on the other. Two poles of one geography of fear.[150] Everyone ends up in prison, behind walls, gates, locked doors, bars. The only differences (and of course these are not trivial) are that amenities inside privileged, quarantined spaces are far more sumptuous, and that the folks at the top of the social hierarchy lock themselves in *voluntarily.*

Moving to the level of society as a whole, the history of suburbanization is a textbook example of what positive feedback loops (aka vicious circles) can accomplish in mere decades. Mass flight to the suburbs, flight from undesirable or threatening urban conditions, contributed to the subsequent further deterioration of the cities. As the cities, and society as a whole, grew more scary, another, more intense round of inverted quarantine began, with millions retreating to exurbia, into gated communities, or into the "fortified" style of city living.

This history has taken us a long way down the road toward a dystopian future in which the world is starkly divided between safe and unsafe places, regulated by walls, gates, forts, bunkers—social fallout shelters of all kinds. From the point of view of citizens who value an open society, democracy, and civic virtue, such a future is distressing to contemplate.

Patchy, segregated social life also seems to have disabled the most important corrective mechanism modern society has at its disposal for dealing with serious structural problems. The polarization of social space has created a profoundly divided nation, less and less able or willing to address pressing social problems. As those able to do so retreat from the social worlds of their less fortunate compatriots, much of the urgency to deal substantively with urban problems, the plight of the poor, and so on, drains away. Without a national consensus that something must be done, majorities in Congress will not support the kind of broad social policy initiatives that might ameliorate the impacts of, if not actually address, the underlying causes of dire social conditions.

From the societal point of view, such downward spiraling positive feedback loops have deeply troubling implications. And in the longer run they also turn out to be bad news for every individual. If mass practice of inverted quarantine contributes to further worsening societal conditions, people will be motivated by those deteriorating conditions to try even harder to protect themselves. They will resort to ever more, and more extreme, forms of inverted quarantine.

Their retreat deeper into personal protective shells can only further erode society's will and ability to deal with the situation. Inevitably, for those who can afford it, inverted quarantine must become ever more the central organizing principle of everyday life. At the same time, self-barricading grows ever less effective, requiring more intense exertions to achieve some modicum of protection. In effect, in the long-run, inverted quarantine helps produce the conditions for its own progressive failure as a self-protection strategy—a classic example of the unintended consequence.

II
Assembling a Personal Commodity Bubble for One's Body

Assembling a
Personal Commodity
Bubble for One's Body

Many Americans fear that they are constantly exposed to toxics in their immediate environment. They think that tap water is contaminated with chemicals and is not safe to drink, that foods have pesticide residues, hormones, and antibiotics in them, that the air carries invisible poisons. Pay no attention, go about your life, eating whatever, drinking, breathing without giving it another thought? It would be foolish to do that.

The response has been swift, even dramatic. Within a couple of decades, bottled water, water filters, and organic foods have gone from being marginal, niche commodities to becoming mass consumer items. Today one finds literally hundreds of other "natural," "green," or "nontoxic" products, household cleaning products, personal hygiene products, cosmetics, clothing, bedding, furniture, and so on, on the market.

Millions of people seem to have embraced the idea that one can fend off these new threats, keep toxics from entering the body, by buying and using such products. I will be examining this most recent development in the use of inverted quarantine methods in Parts II and III of this book.

* * *

I think I should say explicitly at this point that I do not think it is irrational for people to worry about toxics or to believe they are at risk. Our environment does now have many, many hazardous substances circulating in it. These substances *are* in tap water, in food, in the air we breathe, as I will describe in chapters 3, 4, and 5. Biomonitoring studies, a relatively new kind of research that directly examines subjects' bodily fluids, confirm that, indeed, many of these substances are now not only "out there," but they are inside us, too.

Early in the PBS documentary "Trade Secrets," Michael McCally, a physician and administrator at Mount Sinai School of Medicine, takes a blood sample from Bill Moyers. Toward the end of the show, we see McCally tell Moyers the results. Moyers's blood has been tested for 150 industrial chemicals. Eighty-four different chemicals are found, including 31 different PCBs, 13 dioxins, organophosphate pesticides (such as malathion), and organochlorine pesticides (such as DDT). These toxic substances were in Moyers's blood, McCally tells him, because they are *everywhere* in our environment. He says, "You have thirty-one different PCBs . . . [PCBs] are all over the place. . . . you got them into you through the air you breathed. Some of them get down in groundwater. Some of them get coated on food. You didn't get them sort of in one afternoon because you ate a poisoned apple."[1]

The first large biomonitoring study, the Center for Disease Control and Prevention's "National Report on Human Exposure to Environmental Chemicals,"[2] released in 2001, looked for the presence of just twenty-seven substances,[3] but important ones, metals such as lead and mercury, pesticides, phthalates (a plasticizer, not a "household name" like lead or arsenic, but found in many mundane household products, such as soap, shampoo, deodorant, nail polish, soft plastic toys, and *pacifiers*). The study confirmed the presence of many of these substances in Americans' bodies. A reporter showed the results to McCally, the same physician we met in Moyers's documentary. His comment: "What's really remarkable about the CDC's results is that in several instances the chemical exposure levels they measure in the real world are higher than levels predicted by scientific panels, the National Academy of Sciences, and other experts."[4]

The CDC released the "Second National Report" in 2003.[5] The list of chemicals tested for grew from 27 to 116. Of that 116, researchers found 89 in Americans' bodies—PCBs, pesticides, herbicides, phthalates, disinfectants, and others. The encouraging finding was that the amounts of some substances found in people's bodies, such as lead, PCBs, and metabolites of cigarette smoke, had gone down from levels detected in the population at an earlier time. The finding was encouraging because these are substances that are either rigorously regulated or have been banned outright, so their decreasing presence in bodies over time is concrete evidence that regulation does work. Less encouraging was the fact that most of the substances the researchers looked for they indeed found. Also distressing was that the researchers found some contaminants in higher concentrations in children's bodies than in adults'.[6]

In CDC's third report in this series, released in 2005,[7] the number of substances tested for grew to 148; the list included more kinds of pesticides, herbicides, and insecticides, more polycyclic aromatic hydrocarbons (PAHs), and more phthalate metabolites. The overall picture remained much the same. Concentrations of regulated substances, such as lead and the metabolites of cigarette smoke, continued to go down. The substances tested for the first time were, distressingly, found to be widespread in the population. For example, pyrethroid insecticide (used in household roach sprays) was found in 76 percent of the sample; chlorpyrifos (another insecticide), in more than 50 percent. The CDC tested for 23 polycyclic aromatic hydrocarbons for the first time and found all 23 in the population.[8]

The CDC studies are important because they have large sample sizes, thousands of subjects, orders of magnitude larger than some other biomonitoring studies that have been carried out in recent years. Those other, smaller studies,[9] may vary from each other and from the CDC studies in some details, but by and large they all report similar findings: every subject tested has dozens of substances in her or his body, such as PCBs, dioxins, metals, pesticides, fire retardants, and plasticizers.

Once in people's bodies, these substances can be passed on from one generation to the next. They can be passed through the umbilical cord before the infant is born. Umbilical cord blood was collected at the births of ten healthy newborn infants. Researchers then looked for the presence of any of 413 substances in the babies' cord blood. In all, 287 substances were found. Each infant had on average about 200. The cocktail mix of chemicals found in cord blood mirrors what other studies find in adults: PCBs, pesticides, dioxins, furans, PAHs, mercury, perfluorinated chemicals (PFCs such as PFOS, a stain repellant found in ScotchGard, and PFOA, perfluorooctanoic acid, used to make Teflon).[10] A mother's bioaccumulation of toxic substances can also be "downloaded" to her baby when the baby is nursed. Biomonitoring studies have found hundreds of different toxics in breast milk, among them dioxins, PCBs, DDT, flame retardant, and perchlorate (rocket fuel).[11]

These substances are mostly present in our bodies in tiny amounts. We know that many of these substances are carcinogenic or have neurological or reproductive impacts, but because chemicals are typically tested at higher levels of exposure, one chemical at a time, no one knows right now exactly what happens to a person who is exposed to many substances, simultaneously, at low levels, over long periods of time.

Is chronic, low-level, multiple simultaneous exposure harmful? No one can say for sure. Until recently it was generally assumed that the dose-response relationship was monotonic and well behaved. That is, the lower the dose, the less likely there would be adverse health effects. It was also believed that one could, at least theoretically, always find a threshold below which there would be no health effect at all. Recent findings have raised doubts about those assumptions. As research methods have improved, it has often been the case that substances are found to be harmful at lower exposures than previously thought.[12] Effects are now being found at extraordinarily low levels of exposure.[13] It is becoming less certain that "no effects" thresholds really exist. The finding that some substances appear to have bigger health effects at low levels of exposure than they do at higher ones further complicates the picture.[14] What about exposure to many different substances simultaneously? At higher levels of exposure, synergistic effects of exposure to two or three substances at the same time have been documented.[15] Exposure to *hundreds* of substances simultaneously at very low levels? A complete unknown, as far as I, an educated layperson who has looked at some, though far from all, of the literature, can tell.

It is appropriate to recall here that there have been plenty of unpleasant surprises in the past fifty years, as substances not believed to be hazardous when they were first introduced into the environment were discovered in retrospect to have had serious adverse health effects. Given that history, it is not unreasonable for one to conclude that absence of proof (of harm) is not proof of absence, and that it is not irrational, in the absence of definitive evidence one way or the other, to think that it is best to err on the side of caution. Why should one not do what one can to try to keep from ingesting all those toxic substances, even in tiny amounts, letting them bio-accumulate, silently, invisibly, in one's body?

* * *

Even so, that does not mean that buying organic food, wearing "natural" clothing, filtering tap water, and so forth is the answer. Further along, in chapter 6, I will try to determine whether or not buying and using these products works. At this point, I just observe that people feel threatened, that their fears are not baseless, and that millions—actually, tens if not hundreds of millions—are using inverted quarantine–type products, hoping that that will reduce their exposures.

In this part of the book my goal is primarily descriptive. I wish to describe the variety of toxics that threaten to enter the body and the many products that promise to keep them from doing so. There are now so many inverted quarantine products on the market that I began to despair of how to write about them all without getting lost in the details, wearying and losing the reader too.

My solution was to think not of products but instead of the human body. That simplified matters a great deal because there are really only three ways that toxics from the environment can enter the body: when a person eats, drinks, or breathes (through either the lungs or the skin, which can also be thought of, in a way, as "breathing"). An inverted quarantine approach to keeping the body from being contaminated by harmful substances requires that one try to control what does—and what does not—enter the body via one or more of these three *processes of incorporation* (they can also be thought of as portals or points of entry to the body). A chapter is devoted to each of these processes of incorpora-tion: chapter 3 to drinking, chapter 4 to eating, and chapter 5 to breathing. Each chapter describes the dangerous materials that can enter the body via that process or portal, and then describes the products that promise to protect consumers from those threats.

3. Drinking

Sometime during the 1980s, beverage industry analysts began notice that consumption of bottled waters had started to rise, and rise rather dramatically. Bottled waters had been on the market a long time. Immigrants from Europe had brought with them long-held beliefs in the medicinal properties of mineral water, so bottled mineral waters had a small but dedicated following. Local bottlers had also long been making deliveries of five-gallon jugs of water to business offices, supplying those "coolers" around which, allegedly, so much office gossip takes place. But sales of those products did not add up to anything like a significant fraction of beverage consumption. The market was so tiny that when trade journals tabulated how much of various beverages Americans were drinking each year, they did not even bother to cite a number for bottled water.

In 1975, consumption of bottled water reached one gallon per person per year, then kept going up, higher every year.[1] Their interest piqued, industry analysts looked at the figures and saw that most of the growth was happening in just one segment of the market. It was not a sudden demand for more mineral water or seltzer. It was not a sudden jump in "cooler" water orders. The growth was all in one- or two-gallon plastic containers of "still" (i.e., not carbonated) water. They concluded that "a large part of bottled water's sales are a replacement for tap water."[2]

Why is this happening? beverage industry analysts asked themselves. Glass after glass, gallon after gallon, bottled water cost maybe a thousand times as much as tap water. Water is heavy, and it is bulky. It takes some effort to buy it and bring it home. It is not that easy to lift it into the shopping cart or to lug it from the car to the house. Why would people spend the time, money, and effort to do that?

A century ago, the quality of the water supplied to the public for drinking, cooking, and washing was awful, a real health hazard. Outbreaks of waterborne illness were not uncommon. New treatment technologies, especially chlorination, changed that. Drinking water quality improved, and people's trust in the water supply did too. Now, apparently, something new was changing people's perception. It seems that modern environmentalism had not only convinced Americans that the nation's rivers and lakes were polluted, it had also made them suspicious of the water coming from the tap. Bacteria might no longer be a problem, but the water might be contaminated with toxic industrial or agricultural chemicals. *That,* market analysts wrote in the industry's trade journals, was the driving force behind the surge in bottled water consumption: "consumers' growing concern that toxic chemicals may be leaking into their drinking water supplies,"[3] and "Americans' fears, whether perceived or real, that toxic chemicals may be flowing from their faucets . . . continue to grow."[4] In other words, industry analysts believed that the sudden interest in bottled water was, in my terminology, a spontaneous inverted quarantine response.

Recent surveys show that the public's mistrust of tap water has only grown since. The 1999 National Consumer Water Quality Survey found that "about three-quarters [of American adults] have some concern regarding the quality of their household water supply" and "almost half are concerned about possible health-related contaminants." Two years later, a follow-up survey found those numbers had grown bigger still. Eight-six percent agreed they had "concerns about their water," and "51 percent worried about possible health contaminants."[5]

Once the industry understood why consumers were suddenly so interested in bottled water, it began to market its product explicitly as the answer to people's concerns, as a safer alternative than simply drinking water straight from the tap. (Later, marketing campaigns would also sell bottled water as a healthy alternative to alcohol and/or calorie-rich soft drinks and as a convenient source of water for people on the go.) The marketing of bottled water has gone very well. Consumption has grown spectacularly, practically every year, for the past thirty years. Today, bottled water is the nation's second best-selling beverage. It outsells coffee, milk, and beer. Only carbonated soft drinks still sell better.[6]

At the same time, water filters, ranging from professionally installed below-the-counter units to simple tabletop drip filters—clearly another inverted quarantine response to concern about tap water—have also become very popular.

Water Quality in the United States

Is there a problem with our drinking water? I have to admit that until recently I thought there was not. I thought it was kind of paranoid to worry about the safety of tap water. Then I started working in earnest on this chapter. I learned that rivers, lakes, and aquifers, including bodies of water that are sources for drinking water, are quite polluted. I learned that there are serious issues with drinking-water standards. Some standards are too lax. Standards have not been established for some hazardous substances in our nation's waters. Standards are set for one substance at a time, ignoring the fact that a body of water may have dozens of different substances in it. Even if each substance is present at a level that meets the standard, little is known about additive or synergistic effects. Therefore, water that is officially in full compliance with the standards of the Safe Drinking Water Act may still not be safe to drink.

Although it is not possible to be precise or to quantify the risk,[7] I now believe, given the facts I am about to discuss, that it is not irrational for people to be suspicious of the water that flows from the kitchen tap. Obviously, we do not have thousands of people dying simply because they are drinking impure water. The risks are more subtle than that and have to do with the potential adverse health effects of ingesting, over long periods of time, low doses of numerous toxic substances simultaneously.

To better understand the nature of the risk, I will here follow the stream, so to speak, from the pollution of source waters through the partial cleaning of that water by the water treatment infrastructure to the officially clean, but in fact questionable, fluid that comes out of the tap in our homes.

Rivers, Lakes, Groundwater

In its August 1, 1969, issue, *Time* magazine told of a fire on the Cuyahoga, a river that flows through Cleveland, Ohio. "Some river!" the article exclaimed. "Chocolate-brown, oily, bubbling with subsurface gases, it oozes rather than flows. . . . 'The lower Cuyahoga has no visible life, not even low forms such as leeches and sludge worms.'" The article then describes how "the oil-slicked river *burst into flames* and burned with such intensity that two railroad bridges spanning it were nearly destroyed."[8] A river on fire. Water burning! Could there be a more perfect image of Nature fatally out of kilter? In fact, the actual event was not quite that dramatic. A stray spark from a construction project apparently ignited oil-soaked debris snagged on a bridge mooring. But we are dealing with American political culture here. The perception that an issue is real and is important is not created through a sober, steady accumulation of evidence but through a dramatic display of images that convey that we have a PROBLEM. American society had been abusing its waters for decades, treating its rivers and lakes as convenient sinks down which we could pour all sorts of urban, industrial, and agricultural wastes. Various regulatory laws had been passed over the years, but it was a loose patchwork, incomplete and only haphazardly enforced; practically speaking, disposal of materials into the nation's waters was essentially unregulated. Water quality had grown steadily worse. Few seemed to care.

But now there was the Cuyahoga. The article in *Time* fixed the image of the River on Fire in the popular imagination, not as it really happened but—much more effectively—as some sort of medieval vision of Hell. A perceptual threshold had been crossed. It was now official; the United States had a problem with its water.

Congress passed the Clean Water Act (technically, the Federal Water Pollution Control Act) in 1970, shortly after mass media coverage of the Cuyahoga River fire. The act set forth ambitious goals for improving water quality in the United States. All waters in the nation would be fishable and swimmable by 1983; by 1985 pollutants would no longer be discharged into the nation's waterways.

Twenty years after 1985, water quality has improved, but we are nowhere near achieving those goals. The waters are not all fishable and swimmable. Huge quantities of pollutants are still being discharged into the nation's waterways.

The United States has over 2,800,000 miles of rivers and streams. Assessed in 1998 for the Environmental Protection Agency's (EPA) biennial water quality report to Congress, about a third of those river miles, 35 percent, were found to be "impaired for one or more [designated] uses"; another 10 percent were judged to be "good

[but] threatened."[9] The United States has about 41 million acres of lakes, reservoirs, and ponds. Almost half of those lake acres, 45 percent, were found to be "impaired for one or more uses"; another 9 percent, "good [but] threatened."[10] In another publication, the EPA writes that "the overwhelming majority of Americans—over 218 million—live within ten miles of a polluted waterbody."[11]

In addition to surface waters, we also use about 77,500 million gallons of groundwater every day. Groundwater irrigates crops, waters livestock, and provides drinking water for 46 percent of the nation's population (99 percent of the rural population).[12] Less is known about the condition of the nation's aquifers than about the condition of surface waters,[13] but regulators say that groundwater contamination is widespread.[14]

The Clean Water Act has reduced emissions from *point* sources, that is, where water is polluted by one discrete source or by a small number of discrete sources, pipes discharging effluent from several factories, for example. The act has had more difficulty addressing the problem of *non-point* source water pollution, where water quality is degraded by substances coming from a large number of much smaller and/or diffuse sources simultaneously, for instance, when a river runs through a farming community, picking up agricultural runoff, plus runoff from the streets of a nearby town. Even with point sources, progress has been hampered by the fact that the goal has not been to completely eliminate discharges but only to lower the discharges to some level deemed low enough that the waterway can be considered fit for the uses for which it is designated. That means that water quality is better than it would have been if discharges were completely unregulated, but the water never gets completely clean.

Stepping back a bit, it seems to me, though, that the fact that the Clean Water Act has not completely cleaned up the nation's waters has to be explained by something bigger and more fundamental. As I am about to detail, water pollution is the daily, ongoing, inevitable consequence of the way we live in our cities, how we grow our food, and how industry produces things that we consume. Water pollution is *inscribed* in our way of life. The Clean Water Act was meant to improve water quality, meant perhaps to bring it back from some rather gruesome brink before it got completely ruined; regardless of its name, though, the act was never meant to have the power to transform society as radically as it would have to be transformed to make the nation's waters completely clean again. Granted the act has done some good, perhaps lots of good, but the water is still contaminated. Let's review why.

Agriculture

Conventional agriculture uses massive amounts of fertilizers (manure, chemical fertilizers) and pesticides (herbicides, insecticides). Only some of the nutrients in fertilizer are taken up by the crop plants. Pesticides leave residues on the plant and in the soil. When crops are watered, whether by rain or by irrigation, excess fertilizer and pesticides either run off to nearby streams and lakes or percolate down into the local aquifer, especially "during high-flow conditions that result from spring rains, snowmelt, and (or) irrigation."[15] A U.S. Geological Survey (USGS) study of nutrients and pesticides in the nation's waters finds that streams that run through agricultural areas carry heavy loads of "nitrogen, phosphorus, herbicides and insecticides." The study states that "the highest rates of detection for the most heavily used herbicides—atrazine, metolachlor, alachlor, and cyanazine—were found in streams and shallow ground water in agricultural areas."[16]

Two-thirds of the groundwater pumped each day from aquifers, about 50 billion gallons, is used to irrigate crops.[17] A lot of that water then *returns* to the aquifer, now more contaminated. Irrigation "continually flushes nitrate-related compounds from fertilizers into the shallow aquifers along with high levels of chloride, sodium, and other metals." As this cycle repeats, "agricultural fertilizer and pesticide applications . . . [produce] a general deterioration of ground water over a wide area."[18]

Agricultural pollutants also move around in more circuitous ways: "The atmosphere is an important part of the hydrological cycle that can transport nutrients and pesticides from their point of application and deposit them outside the area or basin of interest. . . . Nearly every pesticide that has been investigated has been detected in air, rain, snow, or fog throughout the country at different times of year."[19]

Animal Agriculture

Most of us, these days, live in cities and suburbs, far from real farms. We have in our heads innocent images of barnyards and farm animals, 4H Clubs and the county fair. Anyone who has seen a modern feedlot or one of today's chicken or hog operations knows that what was once quaintly called "animal husbandry" and, more recently, "animal agriculture" is today more akin to industrial production than to any traditional notion of farming.

Animal agriculture affects water quality in two ways. Since keeping the price of meat as low as possible is the whole point of raising cows, pigs, and chickens this way, conventionally raised farm animals are not fed organically grown grain. The impact varies with the way animals are fed. "Range grazing" has the least impact because "land grazed by animals is seldom enhanced by the application of fertilizers or pesticides." In the case of pasture grazing, the impact is greater because "to encourage selected plant species to grow . . . fertilizers or pesticides may be applied."[20] Animals in Concentrated or Confined Animal Feeding Operations are fed mostly corn and soybeans, grown the same way as conventionally grown crops grown for direct human consumption, with repeated applications of pesticides and fertilizers. There the impact on water is similar to the impact one sees with conventionally grown crops meant for direct human consumption, but the impact is greater per pound of food produced because it takes pounds of feed to produce one pound of meat.[21]

In addition, animal agriculture impacts water quality a second way, by generating mountains of animal wastes:

> Nationwide, about 130 times more animal waste [manure, wastewater, urine, bedding, poultry litter, and animal carcasses] is produced than human waste—roughly 5 tons for every U.S. citizen. . . . Animal waste runoff can impair surface water and groundwater by introducing pollutants, such as nutrients (including nitrogen and phosphorous), organic matter, sediments, pathogens (including bacteria and viruses), heavy metals, hormones, antibiotics, and ammonia. These pollutants are transported by rainwater, snowmelt, or irrigation water through or over land surfaces and are eventually deposited in rivers, lakes, and coastal waters or introduced into groundwater.[22]

Industry

American industry generates vast quantities of hazardous wastes. I have seen estimates as high as 2.9 billion tons per year.[23] The fate of some of these wastes is reported to the EPA under the Toxics Release

Inventory (TRI) program. Most wastes are not reported as TRI wastes.[24] TRI does, however, offer a window on what happens to the most toxic of industrial wastes.

When I wrote the first draft of this chapter, 1999 was the latest year for which the EPA had published TRI data. Industries reported that they managed close to 30 billion pounds of TRI wastes that year. Of that, about 10 billion pounds were recycled, about 3.5 billion pounds were incinerated in energy recovery facilities, over 8 billion pounds were "treated," and about 7.5 billion pounds were *released*.[25] Looking at the reporting categories, it seems to me that almost 90 percent of that 7.5 billion pounds were released in ways that could, eventually, result in the wastes ending up in rivers, lakes, or groundwater. A small fraction of the total, 259 million pounds, were discharged directly to surface waters.

The amount of "on-site land releases" was reported as 4.75 billion pounds. Of that, 0.22 billion pounds were disposed in on-site landfills that met Resource Conservation and Recovery Act (RCRA) standards for hazardous waste landfills; over 4.5 billion pounds were disposed otherwise, which means, I think, that wastes were put in the ground in ways less likely to keep them from migrating. Even state-of-the-art hazardous waste landfills, built and operated in full compliance with RCRA standards, leak eventually.[26] Materials deposited in less technologically advanced landfills, into pits or trenches, can easily move off-site, contaminating surface and groundwater. The EPA reports "well-defined, localized plumes . . . emanat[ing from] . . . landfills, waste lagoons, and/or industrial facilities." The EPA cites industrial facilities and the sites where those wastes are—improperly—stored or disposed (landfills, surface impoundments, injection wells, wastepiles) as "major sources" of groundwater contamination.[27] If those contaminants are "mobile [and] persistent," they can travel far. Volatile organic chemicals are widespread in "ambient groundwater . . . [even] in areas where there are *no known point sources* of contamination."[28]

Finally, about 2 billion pounds of TRI wastes were released into the air. Some of those wastes also end up in water. Take mercury as an example. Authorities frequently advise people to not eat fish they catch because the fish are likely to have high levels of mercury.[29] There is mercury in fish, obviously, because there is mercury in the water. Here is the interesting point though: "Industries . . . discharge very little mercury directly into surface waters . . . almost all of the mercury released by permitted polluters . . . enters the air." Mercury released into air is "the most significant source of mercury contamination in surface waters and fish." The EPA says "atmospheric deposition" (that is, of pollutants released into air that eventually come back down) is a "leading source" of lake impairment.[30]

As I completed the latest draft of this chapter, the EPA had just published 2004 TRI data. The total number of pounds of waste released had gone down, to about 4.24 billion pounds. Of that, 0.24 billion pounds were released to surface waters directly, 1.54 billion pounds to land (non-RCRA landfills, surface impoundments, "land treatment"), and 1.55 billion pounds to air.[31]

Urban Wastes

Early modern cities were unspeakably filthy. Towering plumes of black factory smoke merged into a pall overhead. Factory wastes were either shoveled into pits or poured, untreated, into the nearest lake or river. Household wastes piled up in the streets. Bodily wastes and wastewater from bathing and washing clothes were poured into open sewers. The urban environmental reform movement of the late nineteenth and early twentieth centuries successfully fought for improvements. Garbage was collected and hauled to landfills. Sewer systems collected and carried away wastewater.[32] Later, wastewater treatment plants were built; wastewater was filtered and chemically treated before it was discharged back into stream, lake, or ocean. Although some of the worst of urban environmental problems were solved long ago, urban sources still pollute and impair rivers, lakes, and groundwater.[33]

Leaking gasoline station tanks and leaking chemical tanks cause underground plumes of "benzene, toluene, ethylbenzene, xylenes, MTBE."[34] These tanks are found in greatest numbers in urban and suburban areas. Dry cleaning establishments are a major source of VOCs (volatile organic compounds) in groundwater.[35] Insecticides and herbicides are widely used in urban areas, "insecticides . . . in homes, gardens, commercial and public areas[;] herbicides . . . control weeds in lawns and golf courses, and along roads and rights-of-way."[36] Rain washes these chemicals into local streams or carries them down into the aquifer. The USGS reports that some "insecticides are found . . . at higher concentrations, in urban streams than in agricultural streams."[37]

Postconsumer Wastes

Official documents still organize their discussions around traditional distinctions such as industrial, agricultural, and urban. It would be clarifying if they added a fourth category, postconsumer waste, which includes some of what is listed now as "urban" waste but is a broader category defined not so much by geography (since personal consumption is *residential*, it happens wherever people live, shop, and play, be that in suburbs, cities, or small towns) but by its functional position as the last phase of the product life cycle. Our medicine cabinets are filled with prescription drugs, over-the-counter drugs, cosmetics, and personal hygiene products. The food in our pantries and our refrigerators contains all sorts of preservatives, artificial colors, trace amounts of pesticides, and trace amounts of hormones and antibiotics fed to farm animals. In bathrooms, kitchens, and garages, one finds all sorts of cleaning products. All over the house, furniture, electronic equipment, and household items of all kinds are slowly wearing out. Once we are done using an item, we throw it away or pour it down the drain.

Postconsumer waste thrown away as "garbage" is trucked to a landfill. Every landfill, even a well-designed, state-of-the-art landfill, eventually leaks.[38] The EPA says that leachate (the generic term for the fluid that leaks from landfills) carries heavy loads of bacteria, hazardous wastes, organic materials, and sediment.[39] Postconsumer waste that is poured down the drain is piped to a wastewater treatment plant. Wastewater treatment does not remove all harmful substances. Some types of organic chemical wastes, such as hormones, antibiotics, fire retardants, and prescription drugs, "pass through sewage-treatment plants virtually untreated."[40] Treated water "may still contain toxic chemicals, nutrients, and other pollutants."[41]

Because landfills leak and because treatment plants do not remove all the impurities in wastewater, postconsumer wastes find their way into groundwater, rivers, and lakes. In a pioneering study, researchers from the USGS took samples from 139 rivers in thirty states and tested for pharmaceuticals, hormones, and other postconsumer contaminants.[42] They tested for ninety-five different substances and found eighty-two of them, a variety of antibiotics, several types of steroids, reproductive hormones, prescription drugs (analgesics, blood pressure medicines, antidepressants), nonprescription drugs (acetaminophen, ibuprofen), deodorizers, fragrance, plasticizers, detergents, antimicrobial disinfectant, and fire retardant. "The detection of multiple OWCs (Organic Wastewater Contaminants) was common for this study, with a median of seven and as many as 38 OWCs being found in a given water sample."[43]

In another study, water drawn from wells around New Jersey was found to contain hundreds of volatile and semivolatile organic chemicals. More than 20 substances were detected in thirteen of the twenty-one wells; one well had more than 160.[44] Brian Buckley, a coauthor of the New Jersey wells study, wrote this in a newspaper article:

> fire retardants . . . plastics . . . fragrances and flavors . . . the medicines and food additives we consume every day enrich our quality of life and keep us healthy. But they also end up in our water. All these chemicals—antidepressants, caffeine, birth-control pills, deodorants and thousands of other chemicals—are released in small amounts from millions of toilets and drains from homes all over the country.

Buckley concludes,

> It's a common bit of wisdom—You are what you eat. It's just as true that you are what you drink, or, as recent studies are telling us—you drink what you are.[45]

Water: Our Society's Portrait of Dorian Gray

As I read all these reports and studies, I kept remembering Oscar Wilde's tale *The Picture of Dorian Gray*. Dorian, handsome, charismatic, leads a profligate life of ever-greater sin and excess but shows none of the effects outwardly. He stays dashing, fit, charming, and desirable. Meanwhile, hidden in a closet, a portrait of Dorian ages for him and faithfully displays all the effects of a life of physical and moral debauchery. Dorian stays young and handsome while by the end of the book his painted double has grown "withered, wrinkled, and loathsome of visage." Not to be too melodramatic about it, but it strikes me that the portrait of Dorian Gray may be an apt metaphor for the relationship between the American economy, the American "way of life," and the condition of our rivers, lakes, and aquifers.

So why has the Clean Water Act had only limited success? Permit me to repeat the observation with which I started this section:

Water pollution is the daily, ongoing, inevitable, normal consequence of the way we live in our cities, how we grow our food, and
how industry produces things that we consume. Water pollution is
inscribed in our way of life. The Clean Water Act has helped some,
especially in identifying and controlling point source emissions, but
the law was simply never designed to change things radically.

America's rivers, lakes, and aquifers have not once again become
what they once were, simply bodies of water. It is more accurate to
think of them not as water but more like *soup,* complex, if exceedingly thin, soup. The list of ingredients varies from place to place.
In one river or lake, the list contains mostly chemicals from local
industry; in another, maybe mostly agricultural chemicals; in another, contaminants leaking from a landfill; most places, a variety
of postconsumer waste chemicals; everywhere, a complex mix of
substances.

Drinking Water

Still, the EPA says that 87 percent of the rivers and streams and 82
percent of the lakes designated as sources for drinking water meet
applicable standards.[46] What does that mean? Not that you can
actually drink from those rivers or lakes. Every hiker and back-
packer knows that you cannot drink the water from a backcountry
stream, even in a wilderness area, without first boiling it, filtering
it, or treating it with iodine pills.

OK, no one claims one can actually drink from source waters.
"There is no expectation that a municipal water source will be
pure. This is the purpose of a water treatment facility."[47] The EPA
is not saying that source water is safe to drink. It is saying that
waters that meet the standard for source water—are, in the EPA's
terminology, "fully supporting" for use as a source for drinking
water—are clean enough already that once they are treated, with
the usual array of water treatments, "flocculation, sedimentation,
filtration, adsorption, and disinfection,"[48] they will meet the standards of the Safe Drinking Water Act (SDWA). *After* treatment,
the EPA assures us, those waters will be safe to drink.

Most tap water does meet SDWA standards. True, during
any one year, about one in ten water treatment works are in non-
compliance, at least for some period of time.[49] Also true is that the
pipes that carry water from a treatment facility to people's kitchens
are aging. Over time, mineral deposits build up in the pipes. These

deposits create an environment in which bacteria can breed, compromising the quality of the water that flows to the home.[50] Still, the system works well, mostly.[51] In the world of American environmental regulation, 90 percent is a pretty high rate of compliance.

Does that mean that drinking water—tap water—is actually safe? That depends, really, on the adequacy of SDWA standards. If standards are too lax or if there are hazardous substances in source waters that are not regulated, drinking water can be formally in compliance with SDWA, can be officially safe, and still be a problem. The question, "how good is that water?" really comes down to the question, "how good are those standards?"

Three Issues Concerning Standards

Under the authority of the Safe Drinking Water Act, the EPA currently enforces about ninety standards for a wide range of contaminants, bacteria, metals, inorganic chemicals, and organic chemicals.[52] These standards are substantial; drinking water quality would be much more of an issue if these standards were not in place and assiduously enforced. A review of the relevant literature, however, raises three important issues: existing standards may not be stringent enough; there are no standards, at present, for many substances known to be in source waters; and the very logic of standard setting, regulating one substance at a time, may be flawed.

Existing standards may be too lax. When a substance is already regulated, new research can show that the current standard is not stringent enough.[53] Let's look at one example. In 1975, the standard for lead exposure stood at 30 mcg/dl. Subsequent research led the Centers for Disease Control and Prevention (CDC) to lower the standard, first to 25 mcg/dl in 1985, then to 10 mcg/dl in 1991. Two recent studies have found, though, that "a rise in lead levels from 1 mcg/dl to 10 was associated with a 7.4-point drop in [children's] IQ" and that "a blood level of 3 mcg/dl . . . delay[s] . . . the onset of puberty" in girls.[54] These findings are prompting calls for further reductions in children's exposure to lead. Here is a hazard that people have known about for at least two thousand years, and we still have not determined a safe exposure level for it.

When Congress reauthorized the SDWA in 1996, it directed the EPA to review and, if necessary, revise all SDWA standards every six years.[55] Unfortunately, the U.S. General Accounting Office

(GAO) reported in 1999 that the EPA had done almost nothing to implement this provision of the 1996 amendments.[56]

Proposals to strengthen existing standards can be very controversial. Tougher standards cost more, and the industries who would bear the cost almost always oppose adoption of the more stringent standard. Consider the recent controversy over the arsenic standard.[57] The current standard, 50 parts per billion (ppb), dates from 1942. Since then, research has linked arsenic in drinking water to "cancer of the skin, lungs, bladder and prostate . . . diabetes, cardiovascular disease, anemia, and disorders of the immune, nervous and reproductive systems."[58] The World Health Organization and many European nations had lowered their standard to 10 ppb. The National Academy of Sciences weighed in, calling for an immediate reduction of the U.S. standard. In its waning days, the Clinton administration lowered the standard to 10 ppb. There were protests from the mining industry (mine tailings are a big source of arsenic in the nation's waters), the wood products industry (arsenic is used in "pressure treating," a process used to preserve wood products), and the Western Governors' Conference. The Bush administration withdrew the new standard in March 2001, saying more studies were needed.

Eventually, the Bush administration relented. Withdrawal of the arsenic standard threatened to turn into a public relations fiasco. Critics were making arsenic a powerful symbol of the administration's disdain for all environmental protections. Everyone knows that arsenic is a terrible poison. The administration is so ideologically driven that it is unwilling even to reduce the amount of *arsenic* in *drinking water*! The 10 ppb standard was hastily reinstated.

The trajectory of the arsenic standard is instructive, paradoxically, because arsenic is not a typical example of other efforts to strengthen standards. In most cases, the substance in question is far less well known, so there is not that kind of media spotlight on a proposed standard for it, and little political risk for those who oppose the adoption of a more stringent standard.

The second issue relating to standards is the need for more of them. The nation's waters have suspended in them many, many substances that are currently not regulated by the SDWA. The USGS's report on nutrients and pesticides in rivers states, "water-quality standards and guidelines have been established for only about one-half of the pesticides measures in NAWQA water samples."[59] The USGS's report on volatile organic compounds in groundwater says, "there were 14 VOCs that were detected that lack drinking water criteria."[60] And recall all the kinds of postconsumer wastes found in rivers by USGS researchers.[61] If there is no standard for a certain substance, water treatment plants are not required to test for it and are not required to remove it or reduce its concentration in drinking water. The substance continues to be there, but the water is considered to be in full compliance—hence officially safe—even if that substance is thought to cause health problems.[62]

Congress has tried to address this problem. Amending the SDWA in 1986, Congress directed the EPA to develop new standards for heretofore unregulated contaminants, twenty-five new standards every three years—a very ambitious timetable![63] A decade passed; the EPA made little or no headway issuing new standards. So when the act came up for reauthorization again, in 1996, a frustrated Congress said, in effect, OK, we will not expect the EPA to develop new standards at that rate, but we will direct it to make *some* headway. As described by the GAO,

> EPA was required to publish, by February, 1998, a list of high-priority contaminants not currently regulated. . . . [the "Contaminant Candidate List," or CCL] . . . Beginning in August 2001 (and in 5-year cycles thereafter), the amendments require EPA to determine whether to regulate at least five of the contaminants on the list.[64]

The EPA did almost meet its first statutory deadline, issuing the first Contaminant Candidate List only a month late. The list was substantial. It named sixty contaminants known to be widespread in the nation's waters and considered potentially hazardous.[65] Soon, though, the GAO was warning that the EPA was not making progress toward actually setting new standards:

> EPA should be conducting research on these contaminants now so that the regulatory determinations and rulemakings associated with these contaminants will be supported by sound science. However, this research is just now beginning for the most part. . . . little or no health effects research has been initiated for the contaminants on the list. . . . epidemiological studies, in particular, can take 4 or more years to plan and conduct.[66]

In 2003 the EPA removed nine of these contaminants from the list. In 2005 the EPA announced a rather leisurely timetable, saying it would keep gathering data on the fifty-one contaminants remaining on the list for some years, perhaps until 2010, at which point the EPA said it would have enough data to determine whether to propose to regulate some of those contaminants.[67]

If the EPA does eventually propose to regulate any of these contaminants, that proposal is almost certain to be opposed. Any weakness in the scientific underpinnings of a proposed new standard will be exploited to challenge and slow the pace of standard setting. As the GAO put it, if the EPA hopes to succeed, its proposals will have to based on "sound science."[68] Getting data that are so good they are practically irrefutable can take years. So if, in fact, any of the contaminants listed on the CCL, or any of the much larger number of substances known to be in water but not listed on the CCL, do turn out to be harmful, it seems pretty certain that exposure to these substances will not be reduced for years to come.

There is, finally, an issue that transcends the question of the adequacy of any particular standard, or the lack of one, for any specific substance. The whole way regulators have traditionally gone about setting standards, seeking to determine a safe level of exposure for each substance individually, may not be at all appropriate for dealing with the actual situation, that is, the constant exposure to tiny amounts of many different substances simultaneously, day in and day out, throughout people's lives.

Study after study is now teaching us that bodies of water in the United States often have in them not one, not a few, but many kinds of chemical substances. I have already cited a study that found rivers contaminated by a variety of postconsumer waste substances, sometimes up to several dozen in the same sample, and another study that found hundreds of volatile and semivolatile organic chemicals in wells in New Jersey, often twenty or more in a single well. An earlier study by the USGS of pesticides and nutrients in water found that

> individual pesticides seldom occurred alone. . . . Streams and ground water in basins with significant agricultural or urban development . . . almost always contain complex mixtures of nutrients and pesticides. . . . Almost every sample from streams and about one-half of samples from wells with a detectable pesticide contained two

or more pesticides. . . . More than one-half of all stream samples contained five or more pesticides, and nearly one-quarter of groundwater samples contained two or more.[69]

Paul Squillace and colleagues' study of volatile organic chemicals in groundwater found, similarly, that "co-occurring VOCs are common. . . . 29 percent of the wells in urban areas had two or more VOCs, and in rural areas, 6 percent of the wells had two or more VOCs."[70]

Today, the typical river or aquifer that serves as a source for drinking water typically contains many different substances. If there is a standard for a substance, water treatment will remove some of that substance from the water but will not remove it completely. Some of it will remain in the water when it is declared officially fit to drink and sent on to people's homes.[71] Add, then, all the substances that are in source waters that do not have to be removed because there are no standards for them. One can only conclude that not only do source waters have numerous substances in them, but drinking water has them too (though, again, in tiny amounts).

Yet standards continue to be set based on the assumption that any one substance's health impacts can be determined individually and a standard for that substance can be set, again individually, without considering the presence of other substances. As Squillace and colleagues point out, at present what is considered a safe level of exposure to any one VOC "do[es] not consider the co-occurrence of VOCs or the co-occurrence of VOCs with other potential contaminants, such as pesticides or nitrate."[72] He could have said the co-occurrence of VOCs, pesticides, nitrate, industrial chemicals, metals, phthalate plasticizers, prescription drugs, nonprescription drugs, hormones, preservatives, antibiotics, disinfectants, detergents, fire retardants, chemicals used in waterproofing clothes and to make nonstick surfaces for pots and pans, perchlorate (by-product of production of rocket fuel), and on and on.

Individually, each exposure may be considered safe, according to the standard set for that one substance. What happens when a body ingests dozens of toxic chemicals at the same time, each in official safe amounts or less?[73] Are the effects additive? Could the interactions among some of the contaminants be synergistic, so that although each of them is present in what appear to be safe amounts, when present together they can cause trouble?

The answer is, we do not know. With so many substances present in a glass of water, in people's bodies, any number of interactions may be occurring. A few have been identified: The "combination of aldicarb, atrazine, and nitrate, . . . the most common contaminants detected in ground water, . . . can influence the immune and endocrine systems as well as affect neurological health."[74] However, only a few of the many possible combinations have ever been studied.[75]

Scientists and regulators should have, but have not, developed a new paradigm for research and standard setting, a paradigm that more closely models the actual way people are exposed to toxic hazards. Standards continue to be set one substance at a time. That may mean that water officially considered safe may not actually be safe.

Summary: Water Quality

What can we conclude about the quality or safety of tap water today? First, we know that *source* waters, the waters that will eventually, after treatment, become drinking water, are laced with materials, by-products of industrial production, by-products of modern farming methods, and postconsumer wastes. That by itself would not be a cause for concern *if* water treatment were adequate. Again, as Joshua Barzilay, Winkler Weinberg, and William Eley write, "There is no expectation that a municipal water source will be pure. This is the purpose of a water treatment facility."[76]

If source waters are impure, then, the question about tap water quality is really a question about water treatment. Right now, the nation's water treatment infrastructure seems to be functioning pretty well, though there are reasons to be concerned about its future (I will discuss future issues with infrastructure in chapter 7). The major issue right now is not with infrastructure but with the adequacy of SDWA standards. Standards may be too lax. Standards do not take into account potential synergistic effects. Standards have not been developed for many substances known to be in source waters. Tap water is, therefore, like river water, a thin, complex soup, even when it is in full compliance with SDWA regulations. When people drink from the tap or use tap water for cooking, they are eating this soup, ingesting trace amounts of chemicals, not one substance at a time but many all at once, constantly, day in and day out.

No one knows what impacts chronic exposure to tiny amounts of hazardous substances simultaneously may have. Brian Buckley, the coauthor of the New Jersey wells study, commented to a journalist, "The question is, 'Is this something the body deals with at low levels, metabolizes, and there's no problem? Or is this something that accumulates in the body?' We just do not know. To be honest, we are just starting to deal with the question."[77]

We might be lucky. It may turn out that it was not *that* dangerous to drink tap water. Still, right now, one can see why worry about the safety of drinking water is not completely unreasonable or irrational.

Bottled Water

When consumers began to buy bottled waters in quantity, it did not take the bottled water industry long to figure out why and, therefore, what to do to grow their product. People are increasingly wary of water coming from the tap? People are turning to bottled water because they think it is less likely to be contaminated? There is your marketing plan. Position bottled water as the pure, uncontaminated alternative to tap water.

Marketing Bottled Water as the Answer for People Worried about Tap Water

When bottled water brands make their purity claims explicit, they do so in two completely opposite ways, depending on the *source* of the water used.

Brands of bottled *spring* water stake the claim that their products are pristine on the alleged fact that until their water was collected and bottled, it had been safely *distant*—in time, space, or both—from modern social life and its corrupting influence. When Odwalla, the natural juice company, began to market its own brand of "geothermal natural spring water," it distributed a poster-size flyer to let consumers know in great detail why this was a superior product. This water was pure because it came to us from a time *before industrialization* and spent the intervening millennia far from civilization:

> ANCIENT FRESHNESS™—The Odwalla water you are now meeting fell on the land as rain, snow, and glacial melt 16,000 years ago. When it began its circular journey deep into the earth, ecosystems

were in balance, the air was clear, the land-scape wild and primeval. It carried this prehistoric purity underground, where it has remained totally isolated from environmental changes.—*This water is as pure as the day it fell to Earth 16,000 years ago.*[78]

FIJI Water makes similar claims:

> water that has never been touched by pollution or dirtied by pollution because it was created hundreds of years before the industrial revolution and it's been locked under the earth in an aquifer in Fiji . . . at the very edge of a primitive rainforest, 1,500 miles away from the nearest continent. . . . Far from pollution. Far from acid rain. Far from industrial waste. . . . when it comes to drinking water, "remote" happens to be very, very good.[79]

A journey through space and time can have *healing* power. It can make polluted water pure again if we have the patience to let nature work its magic before we bottle, market, and consume it:

> IT SNOWED ON MT. OLANCHA THIS SPRING. GOOD NEWS FOR CRYSTAL GEYSER DRINKERS IN THE YEAR 2094. Or will it be 2006? Or 2156? The truth is, it's hard to tell exactly how long it takes for our Alpine Spring Water to naturally purify itself by seeping through layers of granite fractures and crystalline sand. We do know, however, that the journey covers roughly (very roughly, we assure you) 8,000 feet. And by the time this rocky descent is over, the spring water that we bottle is as pure, as refreshing, and as perfect as you're likely to find anywhere.[80]

Brands that source their water from the public water system (or from a river or from an aquifer that may not be pristine) must necessarily base their purity claims on an entirely different argument. Their publicity pieces describe regimes of *hypertechnological intervention* that take what has been corrupted and make it pure again. If technology has disturbed nature, the solution lies in the unrelenting application of even more technology, exercising complete technical control or mastery over physical matter.

Big Sur Water Company's description of what it does is a good example of the second sort of purity claim. Big Sur's water first

> goes through a CARBON FILTER . . . then [it is] put through a VAPOR COMPRESSION₍TM₎ PROCESSOR . . . then through a 1 micron . . . PAPER FILTER . . . [finally,] As the water then goes toward the filler, super

oxygen in the form of ozone (O_3) is injected into the water to assure our water will remain in a pure state after it is bottled.[81]

Ionics's Web site similarly describes how the corporation manufactures its product, Aqua Cool[R] Pure Bottled Water:

> The process used to purify and produce Aqua Cool[R] Pure Bottled Water involves the use of The Ionics Toolbox[SM] of technologies [that Toolbox[SM] has in it: Electrodialysis Reversal, Reverse Osmosis, Ultrafiltration, Adsorption] to obtain the complete removal of all dissolved and undissolved materials from the source water. The resulting highly pure water is then remineralized with a specific "menu" of minerals selected for taste and fortification.[82]

This text is set next to a photograph of a little girl drinking a glass of water. The girl and the glass both appear to be inside a clear bead of water. The bead in turn sits on a deep blue surface amid other beads, like rainwater beading on a freshly waxed car. A sweet young child, inside a bubble, drinking a glass of water, presumably a glass of Ionics's Aqua Cool[R] Pure Bottled Water—the sunniest and most appealing image of inverted quarantine I have yet found.

Few consumers read bottlers' promotional handouts or surf bottlers' Web sites when making purchasing decisions. They see TV and magazine ads. Mostly, they push their shopping carts down the aisle and see the labels on the bottles. A handful of images, a few words—these are the signifiers that are asked to do the bulk of the work of making purity claims. We will not require the services of an assistant professor steeped in semiotic theory to make sense of the messages in the ads and on the labels; all you need to do is wander down the aisle of a supermarket and look. Those labels contain minor variations on a common theme, words and images repeated time and again. The text has been reduced to a bare minimum: "natural," "alpine," "spring water," "pure," "purified." The heavy lifting is done by the image signifiers: rugged mountains capped by snow and ice, evergreen forests, waterfalls, sparkling rivers, appealing shades of blue and green everywhere. Such landscapes adorn even bottles whose contents, the small print on the label discloses, originally came not from a spring high up in the mountains but from some large city's municipal water system.[83] These images do not just catch the shoppers' eyes, increasing shelf appeal, they also constitute a distinct sign system, proclaiming that the contents are pure, pristine, untainted by human presence. Drinking this water is not just better for you than drinking tap water that may be contaminated. Drinking this water magically transports the consumer back, back, long before modern industry polluted everything, back to the primordial purity of wilderness, back to Eden.

There is, finally, what I believe is the most powerful signifier of all—the bottle itself. This water is *bottled*. Someone took the trouble to isolate it from the rest of the environment, to separate it, to segregate it, bottle it, seal it. It *must* be cleaner, purer, better than water that has *not* been thus singled out, the mundane kind of water that just runs out of the kitchen faucet. It must have been cleaner in the first place; otherwise, why would they have thought it special enough to capture it, separate it, *give it a name?*

Such "thoughts" are of course not thoughts at all. They are, at best, preconscious, the kind of signifying described by Roland Barthes in *Mythologies*.[84] Such meanings do not rise to the level of an articulated, verbalized concept. That is exactly what makes them so convincing, so irrefutable. It is not that hard to look at a label and think, "They are trying to manipulate me with these images." But the bottle is a special kind of signifier. It is not a word, and it is not an image. Here, the object acts as its own sign. In its very muteness (it is, after all, called "still" water), in its very being, in its simple facticity, it makes profound claims for itself.

Expanding the Appeal

Marketing searches for ways to boost sales. The industry believed that bottled water sales began to soar because people were getting worried about the safety of the water flowing from the tap. That is fine, but was there a way to broaden the appeal of bottled water so that demand would continue to grow?

Americans were not just worried about industrial chemicals and environmental illness. They were becoming "health conscious" in other ways, too. They were concerned about their waistline, about drinking too much alcohol, too much coffee, too many soft drinks. Millions were dieting and working out.

The choice to buy bottled water could be redefined. No longer just an alternative to tap water, bottled water could be promoted as part of a healthier lifestyle. If one drank bottled water, one could quench one's thirst without consuming calories and without ingesting a dose of our culture's most popular, though legal, psychoactive drugs. To sell bottled water as a lifestyle choice, advertising associated it with images of desirable social identities, being young, fit, attractive, and active. An Evian ad in *Vanity Fair* is a delightfully clear example of this attempt to broaden and partially redefine bottled water's appeal. Drinking Evian "natural spring water" is simply "another" way "to feel healthy," says the ad. The text is tiny, overwhelmed by the photograph of a beautiful and fit young woman in body-hugging tights, doing her stretches. "Evian," the voice in a television commercial cooed, "your fountain of youth."[85]

In the mid-1990s, the industry initiated another campaign to grow sales, a campaign to convince Americans that proper hydration—which was said to require drinking eight glasses of water a day—was critical to good health.[86] In 2000, a poll commissioned by the International Bottled Water Association (IBWA) asked Americans why they were not drinking enough water. "The reason cited most often . . . is the lack of time or being too busy."[87] It is a mobile society. How do you stay hydrated if you are on the go? Barbara Levine, a researcher at Rockefeller University, one of IBWA's partners in conducting this poll, delivered the punch line: "Keeping a bottle of water at your desk or taking one along on a trip is a convenient way to make sure your body has what it needs to function."[88]

Concern about Tap Water: Still a Powerful Motivator

I am not sure that the hydration campaign really convinced anyone to increase their daily water intake. I do think that convenience—the availability of bottled water in small containers that can be taken along on errands, offered at social gatherings or business meetings—has contributed to demand continuing to rise. Recent polls show, however, that inverted quarantine is still the major force behind demand.

Overwhelmingly, with majorities that range from 75 percent to more than 80 percent, Americans tell pollsters that they are "concerned" or "worried" about tap water.[89]

Why do Americans drink bottled water? Why do they filter tap water? In one poll, conducted in 1998, 69 percent said because of tap water's "taste, smell or color"; 49 percent because of "stories in the news about water pollution"; 41 percent because of "convenience."[90] A poll in 2000 asked respondents, why do you drink bottled water? The most frequent answer was "health"; the second most frequent, "taste."[91] Finally, in 2003, "When Gallup asked respondents why they boiled, filtered, treated tap water, or purchased bottled water, the most frequent responses cited were health related issues (33.3 percent), followed by taste (27.7 percent) and convenience (17.5 percent)."[92]

Respondents' answers on these surveys were constrained by how the questionnaires were constructed, and that, unfortunately, introduced certain ambiguities that make precise interpretation of consumers' motivations somewhat difficult. "Health" can mean "I'm afraid tap water has dangerous toxic chemicals in it." Or it can mean "I'm worried about my weight so I want to drink something that has zero calories." "Taste" is open to different interpretations too. Some consumers just do not like the slight chlorine taste left over from water treatment; others may think that water tasting odd or "wrong" could mean that the water is contaminated.

Nonetheless, the overall message in the polling data is clear enough. Most people drink bottled water (or filter their tap water) either because they actually do not like the quality of tap water or they suspect its quality and worry it could make them ill. Rather than do something politically or collectively to improve the public water supply, they try, individually, to assure themselves a supply of water that they think is safer to drink.

A Spectacularly Successful Product

Just a quarter century ago, bottled water accounted for only a tiny fraction of beverage consumption in the United States, less than a gallon per year per person, insignificant when compared to soft drinks, alcoholic beverages, coffee, tea, and milk. Trade journals did not even bother to post consumption figures for it. Since then, growth in consumption of bottled water has been nothing short of spectacular.[93]

As consumption grew, the industry was itself transformed. Twenty-five years ago the "industry" consisted almost entirely of local and regional brands marketed by small, often family-owned firms. One can get a feel for what the industry was like back then in terms of organizational resources and sophistication by looking at the Bottled Water Association's newsletters. They were slim and amateurish with production values perhaps one step up from the pre-xerox, smudgy, blue mimeographs we had in grade school. (These days, IBWA's publications are printed on high-quality glossy paper; they have eye-grabbing graphics and articles that appear to have been written by professionally trained journalists.)

Beverage and food industry giants that had not previously bothered with bottled water bought into the sector or rushed to introduce their own brands. "Years of double-digit growth is bound to draw new players into any industry. The infusion of both domestic and foreign investors in bottled water is slowly changing the industry's small business, regional nature":[94]

> explosive growth of this beverage segment . . . has encouraged large foreign and domestic companies to acquire some of the industry's leading regional bottlers of water. The growth has also encouraged larger bottled water companies to buy smaller firms in order to increase their geographic distribution. Beginning in the early 1980s and continuing into 1987, acquisition fever hit the bottled water industry.[95]

By 1988, ten companies had acquired and merged their way to having over half the market share.[96] The mergers and acquisitions process continued and perhaps even intensified in the 1990s. By 2000, three corporations, Perrier/Nestle, Danone, and Suntory, had half the market; the other half was shared by over nine hundred brands.[97]

Perrier started importing to the United States in 1907. It remained fairly small, marketing only its own fashionable imported product until the 1980s, when it began to acquire some major domestic brands, such as Poland Spring, Calistoga, and Zephyrhills. By 1987 Perrier was already the nation's biggest bottled water company. It then grew bigger still when it bought BCI Arrowhead (Arrowhead, Great Bear, and Ozarka, the nation's second, fourth, and ninth biggest brands) from Beatrice Foods. With that purchase, Perrier Group's market share reached 23.9 percent. In 1992, Perrier was itself bought up by Nestle, the Swiss food conglomerate. Today Perrier/Nestle sells water under numerous foreign (Perrier, Vittel, San Pellegrino) and domestic (Arrowhead, Calistoga, Deer Park, Great Bear, Ice Mountain, Poland Spring, Oasis, Ozarka, Zephyrhills) labels. By 1999, Perrier/Nestle had 28.9 percent of the U.S. market.[98]

Danone Group, best known for Dannon yogurt, had only its own, rather minor brand of bottled water until it purchased Great Brands of Europe, an acquisition that brought it the high-profile Evian brand. Continuing to grow through acquisition, Danone bought AquaPenn (AquaPenn, Great American, Pure American, and Castle Rock brands) in 1998. Then, in 2000, Danone bought McKesson, at the time the third largest bottled water corporation, a corporation that had four top brands, including Sparkletts and Alhambra. That acquisition elevated Danone to the position of second largest corporation in the sector, with 14.8 percent of the market.[99]

Suntory Water Group, established in 1985, has also grown through acquisitions, especially in the late 1990s, when it bought thirty-two bottled water companies. By 2000, Suntory had 9.2 percent of the bottled water market and was the third largest corporation in the industry.[100]

Booming demand finally attracted the attention of beverage industry giants Pepsi and Coca-Cola. Late to enter, but with tremendous resources, capital to invest, advertising might, a vast distribution network in place, shelf space, and vending machines everywhere, by 2002 their two bottled water brands, Aquafina and Dasani, had taken about 10 percent each of market share.[101]

The popularity of bottled water, then, is reflected not only in consumption figures but also in the organization of the industrial sector itself. No longer just a mass of small firms, after a period of very rapid consolidation the industry is modern, highly monopolized, and dominated by giant food and beverage conglomerates.

Income and Bottled Water Consumption

In the introduction, I did not discuss the social class aspect of inverted quarantine. The case studies in chapters 1 and 2 confirm what should have been obvious: an activity that requires purchase of commodities will have an income/social class dimension to it. The more disposable income a person has, the more likely she or he is to deal with things by *purchasing* goods or services. The higher the income, the more likely that person will choose to buy higher-quality, more expensive versions of those commodity items. Here, and in the chapters that follow, I will consider the income/social class dimension of the inverted quarantine response to environmental threat.

The IBWA posts consumption statistics on the Internet. Interestingly enough, it does not show how consumption breaks down by income, only by geographical location (region, state), age, and gender.[102] There seems to be a certain hesitancy to talk about income differences in consumption. Internal to the business, of course, you have to know such vital facts about the nature of your customers. A Web search turns up lots of studies that analyze bottled water consumption by income. You just have to pay thousands of dollars to see it.

If one cannot afford that, one is left with what market studies one can find in business school libraries and the occasional news article that cites proprietary market data. Those sources confirm that consumption of bottled waters does indeed have a class character. The higher the household's income, the more bottled water consumed. That is true for consumption overall,[103] as well as for bottled water purchased while dining out.[104]

People with money are not only more likely to drink bottled water; they also drink a *higher class* of bottled water. Premium brands such as Perrier, San Pellegrino, and Evian are more expensive than the generic plastic gallon jugs one finds in bulk on supermarket shelves. These brands explicitly position themselves as status products meant to be consumed by people with money. Edward Slade, president and CEO of the Fiji Water company, proclaims his ambition to place Fiji in "the finest four-star restaurants" and to achieve "an incredible level of prestige."[105]

The Simmons Market survey suggests that such class-based positioning works. Income determines not only how much bottled water a person buys but also the brand. One survey asks consumers not only about the total amount of bottled water they buy but also about specific brands. As income rose, consumers tended to increasingly prefer Evian over the less prestigious Arrowhead.[106]

At the very top of the class ladder, we can imagine households where every drop of water is either filtered with top-of-the-line filter technology or prestige-brand bottled water. Unfortunately, we do not have good data about the water consumption practices of the very rich, and for much the same reason we do not know any details about the atomic fallout shelter Nelson Rockefeller had built for himself in his posh New York City townhouse. We have, again, only anecdotes, such as this from the Bottled Water Web site: "Racquel Welch not only drinks Evian, she even washes her hair in it. And Michael Jackson orders 32 cases at a time because he literally bathes in it."[107]

Water Filters

The water filter industry markets a range of products, from tabletop pour-through pitchers with filters (such as Brita and PuR) to comprehensive water purification systems that require substantial plumbing skills to install. These are products whose *only* purpose is to provide an alternative to unfiltered tap water. None of the other reasons people may have for drinking bottled water, such as an alternative to alcohol or soft drinks, or having some water along to keep hydrated while on the go, is in play here. In that sense, use of filters is actually an even purer case of inverted quarantine than bottled water use is.

The Water Quality Association (WQA), the water filter industry association, fields public opinion surveys every couple of years to gauge people's feelings toward tap water. The 1999 National Consumer Water Quality Survey found that "about three-quarters [of the public] have some concern regarding the quality of their household water supply" and "almost half are concerned about possible health-related contaminants."[108] Two years later those numbers had grown; 86 percent had "concerns about their home water supply," and "51 percent worried about possible health contaminants."[109] With each new survey, more folks tell WQA that they have some kind of filter in their homes: 27 percent of respondents in 1995, 32 percent in 1997, 38 percent in 1999, and 41 percent in 2001.[110]

The expected class gradient is present here too. The higher their household income, the more likely people are to have filters installed in their home.[111]

* * *

Industry, agriculture, animal agriculture, urban and suburban collective spaces, and millions of individual households all generate streams of hazardous wastes. Released into the environment, waste chemicals find their way into rivers, lakes, and groundwater. Some of those waters are source waters for our drinking water. Water treatment removes some, but far from all, of these contaminants. What remains—we ingest when we drink.

The biomonitoring studies I cited in the introduction to Part II confirm what many have long suspected, that environmental pollution is not just *out there*; it is also now *in here*, inside our bodies. Drinking water that has hundreds of substances in it is one way pollutants end up in people's bodies. The pollutants may be present in tiny amounts, parts per million or parts per billion, but we drink water day in and day out.

No one knows at this time if ingesting all these substances in tiny amounts over a lifetime is hazardous to human health. Scientific uncertainty notwithstanding, there is widespread belief that ingesting all that cannot be good for you. That is why bottled water is so popular. That is why so many households filter water.

Add together consumption of bottled water and use of household filters; a truly impressive numbers of Americans no longer drink water directly from the public water supply. The WQA reports that by 1999 62 percent of Americans, almost two-thirds, said they drank bottled water, or filtered water in their homes, or both.[112] The National Environmental Education and Training Foundation survey, completed about the same time, found "65 percent of all [Americans] either boil their tap water before drinking it, filter it, or use bottled water in the home."[113] The poll done by Gallup for the EPA in 2003 found that 37 percent "reported using a filtering or treatment device," 74 percent bought and drank bottled water, and 20 percent drank "bottled water exclusively."[114]

In this chapter my goal has been to describe why people feel at risk and to describe how their feeling threatened has fueled the growth of bottled water consumption and the use of water filters in homes. In chapters 6 and 7, I will come back to bottled water and water filters and take up some questions that I did not address here, such as: Do these products work? Is the consumer who uses these products really safer? And does mass flight to bottled water and water filters have other consequences, consequences that the users of these products do not see, do not intend, and may not necessarily want?

4. Eating

The term **health food store** used to evoke images of "shriveled produce with brown spots sold in a tiny store with sawdust strewn on a wooden floor and potted ferns hanging from the rafters."[1] That stereotype might not have been that far-fetched a few years ago. "Health food" was for folks who were a little odd, a bit too health conscious, perhaps a bit hypochondriacal. Today, the organic food market is very different: national chains of attractive upscale organic food markets; organics on sale in mainstream supermarkets; organic labels owned by giant food corporations like Procter & Gamble, General Mills, Heinz, and Dole. Sales in the United States of organic foods and drinks topped $10 billion in 2003 and were expected to approach $15 billion by 2005.[2]

Once, not long ago, food safety was not really a big issue. Then came an increasingly lengthy series of disturbing reports about pesticide residues on apples and on other fruits and vegetables, preservatives and food coloring that can cause cancer, hormones and antibiotics fed to farm animals, genetically modified crops and recombinant bovine growth hormone (rBGH), mad cow disease.

The critique has gone in two directions. One, the critique has addressed the likely *environmental* impacts when each year farmers apply a billion pounds of highly toxic pesticides to their crops and feed twenty-five million pounds of antibiotics to healthy farm animals, and when foreign genes are transplanted into a crop plant's genome and millions of hectare of the new, bioengineered crop are planted. Two, the critique has discussed the potential *public health* impacts when consumers ingest pesticide residues, growth hormone residues, and genetically modified vegetables.

There is currently a great deal of uncertainty and controversy about all this. Are these threats to environment and to public health real, or are activists just overreacting, sounding the alarm without just cause? If the risks are real, it is impossible to know right now with any degree of certainty how serious those risks are.

On the other hand, all those substances biomonitoring studies find in Americans' bodies are undeniably there. They got there *somehow,* and we will be lucky indeed if it turns out that having all those foreign substances in our bodies turns out to do no harm.

In any case, perception has gotten there faster than scientific certainty. People feel threatened. That is why consumption of organic foods is growing at fantastic rates.

In this chapter, I first review the range of concerns that have been raised about how fruits and vegetables are grown and how farm animals are raised here in the United States, and how those practices are said to threaten consumers' health. I then document the growth of the organic food sector. Finally, I use marketing surveys to show that rising consumption is mostly—and over time, increasingly—an expression of the inverted quarantine impulse.

I begin with fruits and vegetables. I do so because fruits and vegetables were the first foods to be grown organically, and they still account for almost half of all sales of organic foods and beverages.[3]

Sources of Potential Hazard in Fruits and Vegetables

Pesticide Residues

In the case of produce, most of the discussion to date has been about pesticide residues. It is obvious why. Crops are routinely sprayed with herbicides, insecticides, and fungicides. Pesticides are applied to crops in large amounts, more than a billion pounds each year.[4] These are very dangerous chemicals, known to cause serious health problems. There has been a lot of media coverage; the American public has been told repeatedly that "conventional" farming, as it is now called by the organic farming movement, grows fruits, vegetables, and cereal grain crops in a toxic environment.

After an agricultural chemical has done the work it was meant to do, some of it decomposes. Some of it washes away when it rains or when the crop is irrigated (and ends up in the soil or in streams, lakes, or aquifers, as we saw in chapter 3). Some residues can remain on the plant or be absorbed by the roots and end up in parts of the plant that will be eaten, the fruit or the leaf. Those residues will still be there when the item goes on sale at the grocery store or supermarket and still be there when those foods are served and eaten.

Federal regulations set standards for how much pesticide residue may remain on or in fruits and vegetables. The U.S. Department of Agriculture (USDA) runs a Pesticide Detection Program that buys thousands of samples of fruits and vegetables in markets and tests them for pesticide residues to see if these standards are being met or exceeded.[5] If residues are present in amounts that are below those set levels, the produce is officially considered safe to eat.

The test results are either reassuring, if one believes the federal government's own analysis of the Pesticide Detection Program data,[6] or they are alarming, if one believes analyses of the same data by consumer groups and advocates of organic farming.[7]

A Food and Drug Administration (FDA) press release states that "of more than 10,000 food samples reported from regulatory monitoring, . . . over the last seven years . . . fewer than 50 [exceeded] . . . tolerances (the highest levels legally allowed) set by the Environmental Protection Agency."[8]

The activists who do secondary analyses on the FDA data emphasize how *ubiquitous* pesticide residues are, and they argue that just because these residues are deemed safe by the regulators does not necessarily mean they are actually safe. The Consumer Union emphasizes in its report that the USDA finds pesticide residues on almost every kind of fruit and vegetable tested. Peaches, apples, peas, grapes, spinach, winter squash, and green beans have especially high levels. Looking at the data another way, Consumer Union reports that "37 different pesticide chemicals were detected in apples by the PDP [USDA's Pesticide Detection Program] . . . more than 20 are found in peaches, pears and spinach."[9] Charles Benbrook writes that 80 percent of the fruits sampled by the Pesticide Detection Program and 75 percent of the vegetables sampled have pesticide residues on them.[10]

Furthermore, Benbrook reports, almost half the fruits and vegetables had two or more residues; 5.5 percent of samples had five or more. Other researchers, analyzing data from the FDA, report that "residues of five or more persistent toxic chemicals in a single food item are not unusual."[11]

These researchers are not reassured by officials' arguments that just because residues are present in amounts that do not exceed current standards, these foods can be considered safe to eat. Their concerns echo issues I raised in chapter 3 when I discussed potential problems with safe drinking water standards. Pesticide residue standards in force today may be found, after further research, to not have been protective enough. Recall the typical trajectory of regulatory standards: as scientific understanding improves, earlier standards prove to have been too lax, and the exposure level deemed to be safe has to be lowered. Even if a standard is adequate for adults, it may not be stringent enough to protect children, who "are much less able than adults to detoxify most pesticides, and are highly vulnerable to endocrine disruptors and developmental neurotoxins."[12] Furthermore, safe exposure levels are set for individual pesticide residues, while the real world situation is that consumers eat many different kinds of foods and, therefore, ingest complex mixtures of pesticide residues. Even if these residues are all individually present in amounts that meet federal standards, and even if each individual standard is adequate (and again, that is not at all certain), little or nothing is known about the health impact of exposure to many different pesticide residues simultaneously, in tiny amounts, over long periods of time.[13]

Other Contaminants in Produce

Eating conventionally grown fruits and vegetables may expose the consumer to threats beyond the obvious ones from the residues of pesticides *intentionally* applied to crops. Plants may be picking up other contaminants from other sources in the soil and in water used to irrigate crops.

Soil may be contaminated with pesticides from earlier applications. Some of these pesticides are very persistent. DDT, dieldrin, and chlordane have been banned for decades, yet they are still found in the soil.[14] Soil may have in it volatile organic chemicals, metals, and other waste products, the detritus of industrial civilization, wandering, uncontrolled, settling here or there.

It takes a lot of water to grow crops. Only some of it falls from the sky. Sixty-three percent of all groundwater pumped each year in the United States and 40 percent of all freshwater appropriated for all uses annually goes to irrigate crops.[15] The water quality studies I reviewed in chapter 3 show in abundant detail that surface waters and groundwaters have many kinds of harmful substances in them— pesticides, nutrients, metals, a cocktail of volatile organic chemicals, and traces of many industrial and postconsumer chemicals.

Some of these substances are filtered out by a plant's roots; others are absorbed and stored in the plant's tissue. Consumer Union's secondary analysis of the USDA's Pesticide Detection data shows that some crops have in them detectable levels of dieldrin and heptachlor epoxide even though both were banned in the 1970s and have not been applied to crops since.[16] K. S. Schafer and S. E. Kegley's review of FDA data showed, similarly, that

> POPs [persistent organic pollutants] residues are present in virtually all categories of foods, including baked goods, fruit, vegetables, meat, poultry, and dairy products. . . . the most commonly found POPs [are] the pesticides DDT [note: also long banned] (and its metabolites, such as DDE) and dieldrin.[17]

Plants roots absorb certain metals, such as selenium and cadmium. In fact, some plants do this so well that scientists are considering using plants to remove these metals from contaminated soil.[18]

Here is some recent news: "A leak from a rocket fuel plant into the groundwater near Henderson, Nev., near Las Vegas, has led to perchlorate contamination in the lower Colorado River."[19] The Colorado provides drinking water for many communities in the Southwest, and much of its waters are used to irrigate crops. Leafy vegetables such as broccoli, cauliflower, and lettuce take up perchlorate. Perchlorate can be harmful to people's health—it "inhibit[s] the thyroid's uptake of iodine"[20]—but no one knows right now exactly how much a person can ingest before that begins to happen.

Banned pesticides, metals, and rocket fuel are just some, a tiny fraction, of all the substances in soil and water today. Which of these are taken up by crops, and which are not? I have not been able to locate anything like a full accounting of the fate of these substances.

Researchers are still in the process of documenting what exactly is in America's waters these days. Much of the data I presented in chapter 3 are quite recent. What happens to those many substances in water when that water is applied to agricultural fields? My sense is that we are just beginning to find out.

Recently, for example, two researchers from Johns Hopkins reported that triclocarban (TCC), a substance used to make anti-septic soap, "survives sewage treatment and is being spread . . . [as] sewage sludge . . . onto farmland and released into water."[21] The researchers estimate that about 60 percent of U.S. waters have TCC in them. TCC degrades into a substance that causes cancer in animals,[22] but no one knows right now if TCC is taken up by plants' roots when it is in the sludge spread on farmland or in the water used to irrigate crops, so we do not know if TCC in soil and in irrigation water is a public health issue or not.

As I said, the data are just trickling in, one substance at a time, and even then are often incomplete. That is, even when it is shown that some substance is in the soil or in the water, it takes further work to figure out if that substance is then taken up by crop plants and how much of it is taken up, and therefore how that substance might eventually affect the health of the person who eats that fruit or vegetable.

Genetically Modified Crops

I recall a time a little more than ten years ago when labs in the Bay Area were beginning to grow genetically modified (GM) crops experimentally. The work proceeded with great caution; the experimental crops were grown, for example, in rooms carefully sealed off from the outside environment, the air pressure inside those rooms kept artificially low so air would flow in, rather than out, in case the containment was accidentally breached.

Much of the world has continued to be leery of GM crops. Some nations have bans on importing them, and others moratoriums on growing them. Here in the United States, they were quickly declared officially safe, so today GM crops grow on about a hundred million acres of farmed land.[23] Eighty-five percent of soybeans and half the corn grown in the United States are genetically modified.[24]

The numbers make it look like there is no issue in the public's mind about GM crops, but that is not the case. The FDA says GM foods are safe to eat:

FDA CONSUMER: Do the new genes, or the proteins they make, have any effect on the people eating them?

DR. HENNEY: No, it doesn't appear so.[25]

Activists say otherwise. They cite studies that suggest possible adverse health effects, but they mostly argue that GM *may* eventually be shown to be hazardous and it is madness to transform agriculture and feed GM foods to people before anyone really knows that it is completely safe to do so.[26]

What about the consumer? In Europe there is widespread distrust and resistance. Activists there call GM foods "Frankenfood," once again evoking the allegory of technological feats escaping our control and returning to wreak havoc on us (recall from chapter 1 how the Frankenstein story was used to express dread of the atomic bomb after Hiroshima). The European Union's (EU's) tight controls on the import of GM crops have prompted the U.S. government to file a case against the EU with the World Trade Organization.[27]

When we look at the polls, the American consumer seems less perturbed and not particularly well informed about the GM issue.[28] Why is that? It could be a sign of tacit acceptance. It could be, on the other hand, a consequence of *invisibility*. How can the consumer know, for example, if the soybeans used in a prepared food are soybeans that have been genetically modified? Industry has successfully fought labeling laws that would require GM ingredients be listed on food product labels. That is partly why the issue is largely invisible where it really matters, day to day, when the consumer is in the supermarket, pulling items off the shelf. Still, one can detect some level of consumer concern. Otherwise, why would Gerber refuse to use GM ingredients in its baby foods, or McDonald's stop using GM potatoes to make its french fries?[29] The Whole Foods organic consumer survey found that 68 percent of those who buy organic say they do so in part because organically grown foods are foods that are not genetically modified.[30]

Sources of Potential Hazard in Meat and Dairy Products

Debate about the safety of eating meat (also drinking milk and eating eggs) from conventionally raised farm animals has focused on hormones and antibiotics administered to healthy farm animals and on concerns about what those animals are being fed. Animals can also ingest substances via the water they drink, but that issue has not (yet) gotten widespread attention.

Hormones and Antibiotics

By one estimate, 70 to 90 percent of beef cattle in the United States are injected with pellets that over time release growth hormones, such as testosterone, progesterone, and estradiol.[31] The animals gain weight faster, so they can be sent to market sooner. Shortening the time between when a calf is born or is purchased and the time the grown cow is ready for slaughter makes good business sense. The grower's investment is recouped faster, and that increases his rate of profit.

Growth hormones fed to animals have not been as big an issue here in the United States as in Europe, except for rBGH, recombinant bovine growth hormone, a bioengineered hormone given to milk cows to increase their milk production. In spite of assurances from the manufacturer, Monsanto, and from federal regulators, quite a few parents readily concluded that milk from rBGH-injected cows may not be good for their kids.[32]

Healthy cows, pigs, and chickens are fed antibiotics, among them penicillin, tetracycline, and erythromycin. Antibiotics are fed to animals to help them grow faster, and as a prophylactic measure against the spread of bacterial infections, an ever-present threat when animals are raised crowded together in filthy conditions. About twenty-five million pounds of antibiotics are fed to healthy farm animals each year, an amount far higher than the three million pounds of antibiotics used to treat bacterial infections in people. It is a startling statistic; annually, about 90 percent of all the antimicrobial medicines are used to "treat" healthy farm animals instead of to treat sick people.[33]

Are all these hormones and antibiotics a problem? Just as with regard to pesticide residues in produce, agricultural interests and government regulators generally say "no," while health activists (and European governments, as we will see presently) think otherwise.

The FDA's official position is that giving cows growth hormones does no harm to consumers.[34] European science panels who reviewed the available evidence in 1999 and again in 2000 disagreed. Although "the current state of the art does not allow a reliable quantitative risk assessment," they wrote, the available scientific evidence convincingly shows that "increased exposure to hormones can be associated with an increased risk of cancer and detrimental effects in development." The report went on to say that even if "most of these effects have been demonstrated [only] following high dose exposure, . . . there is not compelling evidence suggesting that these effects do not also occur at low doses."[35]

Considering the "possible consequences of continuous daily

exposure—even to low levels of hormones—to all segments of the human population, including at the most susceptible periods (*in utero* and prepubertal),"[36] the EU banned the use of growth hormones in cattle and banned the import of beef products from animals that had been fed growth hormones. The EU feels strongly enough about protecting the European consumer from the potential impacts of growth hormones in beef that it continues the ban in spite of vigorous protests from American cattle growers and legal challenges brought before the World Trade Organization (WTO) by the United States and Canada.[37]

In the case of antibiotics, the public health issue is not that these pharmaceuticals will remain in the flesh of the farm animals, eventually to be passed on to the bodies of consumers. Rather, the concern is that feeding animals all those antibiotics exerts powerful evolutionary pressures on bacteria, creating ideal conditions for the rapid emergence of bacterial strains that are resistant to most currently available antibiotics, with potentially disastrous consequences for public health.[38]

Animal Feed

There are numerous issues regarding animal feed, the most prominent being the possibility that pesticide residues on crops fed to farm animals are still there when the meat from these animals is eaten, and that feed made from rendered animal by-products can spread bovine spongiform encephalopathy (BSE), or mad cow disease. There are other issues, too, though they have not been discussed as widely.

Pesticide Residues

Corn and soybeans make up the bulk of commercial animal feeds, somewhere between 70 and 90 percent.[39] Since over 99 percent of all corn and all soybeans in the United States are grown conventionally, not organically,[40] the feed made from these crops may have pesticide residues in them. Once eaten, those residues might even bioaccumulate over a lifetime of feeding, especially in animals' fatty tissue.

It is controversial, of course. A beef industry Web site, www .beef.org, ridicules the idea that pesticide residues in meat might be a problem. Those residues, the Web page asserts, break down and/ or are excreted long before that animal goes to market.[41]

Maybe so, but if farm animals ingest pesticide residues either from their feed or from their water, why wouldn't those residues accumulate in the animals' fatty tissues and still be there when the meat reaches market? Persistent Organic Pollutants (POPs) in ocean water bioaccumulate and bioconcentrate in the arctic food chain and are present in high concentrations in seal fat. That is why Inuit women, whose diets include seal, have such high levels of PCBs in their breast milk.[42]

The USDA seems to think that meat can still have pesticide residues in it when it reaches the market. The agency has regulatory standards that specify acceptable levels for pesticide residues in meat.[43] Why would the USDA bother to publish standards if it thought pesticide residues in meat were not an issue? The EPA also seems to think it is an issue. The EPA advises consumers to "trim fat from meat, and fat and skin from poultry [because] residues of some pesticides concentrate in animal fat."[44]

Other Residues in Feed Crops

Earlier I discussed the possibility that crop plants can absorb other contaminants that are in the soil and in irrigation waters. As I have pointed out both above and in chapter 3, many kinds of industrial chemicals and also postconsumer wastes can and do end up in soil and in water. If any of these contaminants are present where feed crops are grown and if the particular crop grown there tends to absorb those contaminants, they can end up in animal feed. These other possible contaminants via feed are discussed much less in the literature than the problem of pesticides, but I have seen reference to some, specifically polychlorinated biphenyls and heavy metals.[45]

Genetically Modified Corn and Soybeans as Animal Feed

Commercially mass-produced feed for cows and for chickens consists mostly (70 to 90 percent) of corn and soybeans.[46] Eighty-five percent of soybeans and more than half the corn grown in the United States are now genetically modified varieties of those two crops. Those two facts considered together must mean that the typical farm animal eats GM feed, quite a bit of GM feed.[47]

Animal scientists and the agricultural biotech sector publicize studies that show GM feed to be safe. Most of these studies find that GM feed does not harm the animals themselves. They grow as well and produce as much milk or as many eggs as animals fed non-GM feed. Somewhat more to the point for us, other studies show that GM proteins or GM DNA fragments *are not* detected in poultry meat or in eggs when chickens are fed GM crops.[48]

Europeans are somewhat more cautious. The European Community has in place a complex permitting system that requires extensive testing of each GM crop before it is allowed to be used as animal feed. The EC has allowed the use of a small number of GM crops, certain varieties of corn, rape, and soybean.[49]

Activists express concern that not enough testing has been done. Problems too subtle to have been detected by the tests done so far may be present, but because agriculture and governments have rushed to adopt the new biotechnology products, millions (no, billions) will have been exposed to the hazards before those problems come to light.[50] Embracing a precautionary outlook, organic meat advocates say organic is safer not only because the feed given to organically raised animals is pesticide free but also because the feed is not made of genetically modified grain.

On the other hand, and unlike some other issues I have discussed here, the American public does not seem particularly concerned about this facet of the GM issue. Americans do seem very uneasy when they hear that farm animals themselves might someday be bioengineered,[51] but they do not—or do not yet—express any great concern about cows and chickens eating GM feed.

Feed Made from Animal Parts and Animal Wastes

Now here is a feed issue that *has* successfully crossed the perceptual threshold. Consumers are aware and are concerned that billions of pounds of meat, animal fat, bone, and blood—all waste products from the slaughterhouse—plus chicken feathers and poultry litter are processed into animal feed.[52]

Because the practice was largely invisible to the typical urban food consumer, most people probably had no idea that rendered animal parts and animal wastes were being turned into animal feed until the appearance of bovine spongiform encephalopathy (BSE), better known colloquially as mad cow disease. Mad cow disease is thought to be caused by aberrant, "wrongly folded," prions, and it is thought that the disease is transmitted from sick to healthy animals when sick animals are slaughtered and the unused parts of their carcasses are rendered and then fed to healthy animals.

Mad cow disease is the issue most in the public's mind when feed made of rendered animal parts is discussed, but it may not be the only issue. Any substance that is stored in fatty tissue, such as di-oxins and PCBs, can be "transmitted" in the same way wrongly folded prions are transmitted. Persistent toxic substances in slaughtered farm animals' unused fatty parts are still there after those parts have been rendered and processed into feed and are, in effect, "recycled" into the bodies of live animals.[53]

Other Feed Additives

Commercial animal feed mostly consists of corn and soybeans, then rendered and processed animal by-products and wastes. The makers of these commercial feeds also add other ingredients, such as clay and minerals. These additives also can be contaminated. I have seen references to the FDA stopping distribution of feed when it found the mineral added to it was contaminated with dioxin.[54] Consumer Reports mentions that clay added to the feed mix can also be contaminated with dioxin and writes that "in fiscal year 2003, dioxin contamination led the FDA to recall 479 feed products from 17 companies."[55]

Animals' Drinking Water

Farm animals do not drink tap water; that is, they do not drink water that has to meet SDWA standards. Farm animals drink river water, pond water, and groundwater pumped from aquifers. As we saw in chapter 3, those waters are often contaminated with a variety of hazardous substances, many different ones at the same time.

Unlike plants, whose roots are selective about the substances they will absorb when they take up water, animals drink it all. When they are thirsty, they just gulp it down, all of it, with whatever chemicals, pesticides, nutrients, volatile organic chemicals, metals, and so on, the water has in it.

What is the fate of the substances in animals' drinking water? Are contaminants broken down or excreted? Do they instead bioaccumulate and are still there when the animal ends up on some consumer's breakfast, lunch, or dinner plate?

Seeking an answer, I consulted several reports written by university-based extension specialists.[56] I found that when these experts discuss the problem of poor-quality feed water, they do so primarily in the context of concern that poor water quality can harm a farmer's income. If animals are offered polluted, poor-quality water, they may stop drinking, become listless, stop growing, fall ill, and even die. That is not good for business. Polluted water as a potential health issue for consumers gets markedly more perfunctory treatment. Nonetheless, several reports make reference, albeit in passing, to the possibility that all sorts of pollutants—pesticides, hydrocarbons, hundreds of different volatile organic compounds, metals, arsenic, mercury, selenium, cadmium, "many industrial chemicals"[57]—can be found in animals' drinking water, and if these substances are present in sufficient quantities, they could "accumulate in the meat, milk or eggs making them unsafe for human consumption."[58]

The National Academy of Sciences recommends maximum contaminant levels (MCLs) for some substances in farm animals' drinking water, but these are only recommendations, not enforceable standards (such as the drinking water standards promulgated by the EPA under the authority of the Safe Drinking Water Act), and they are clearly written with the health of the farm animal, not the safety of the eventual consumer, in mind.[59]

Substances Added When Foods Are Processed

So far I have been describing hazards introduced in the production phase, while growing crops and raising animals for food. Some foods move from farm to market without further processing, but many undergo extensive processing before they are offered to the consumer. According to the FDA, "over 2000 substances [are] directly added to food."[60] These substances are added mostly to "enhance flavor, ensure desirable texture, or retard spoilage."[61] Some additives were banned after they were shown to have serious adverse health impacts. The FDA considers additives still allowed to be used today generally safe; still, its Web site acknowledges problems with some food colorings, fat substitutes, preservatives, artificial sweeteners, and sulfites.[62] The Center for Science in the Public Interest's (CSPI's) Web site is much more categorical in distinguishing what additives it considers safe, versus additives it believes consumers should try to use only in moderation, versus additives that consumers "should avoid [because they are] unsafe in amounts consumed or [are] very poorly tested and not worth any risk." CSPI's "caution; try to avoid" list includes the widely used preservatives BHA and BHT. Its "Avoid; additive is unsafe in the amounts consumed or is very poorly tested" list includes several sweeteners, including aspartame (Equal, NutraSweet), several food colors, and the preservatives sodium nitrate and sodium nitrite.[63]

So, are all those pesticide residues, hormones and antibiotics, and substances added in processing a problem or not? Given the current level of scientific knowledge, no one can say for sure. Generally, federal agencies and industrial groups say there is no problem. Public health activists, politically progressive scientists, and public interest groups think otherwise.

Increasingly, the public seems to be siding with the critics. They are "voting with their feet," or, more aptly, with their pocketbooks. They are buying organic.

The Growth and Mainstreaming of Organic Food

In 1990, Americans were spending about 1 billion dollars a year on organic foods. Since then, sales have grown by about 20 percent a year. Through the magic of "compound interest," 1 billion in 1990 became 2.5 billion by 1995 and between 6 and 7 billion by 2000, and was predicted to reach about 14.5 billion by 2005.[64] By any measure, it is clear that the sector is not just growing; it is diversifying, maturing, and going mainstream.

The sheer number of different organic food items now offered for sale has grown enormously. In the beginning, "organic food" meant organic fruits and vegetables. Produce still leads sales today, but its share of the organic market has fallen below 50 percent.[65] You can now buy all sorts of organic staples: breads and cereals, meat, poultry, fish, dairy products, and baby food. The Organic Trade Association says that "the dynamic organic food sector [is now introducing] 1,500 new organic products each year."[66] Once introduced, some of these items flew off the shelves. From 2002 to 2003, sales of organic meat, fish, and poultry grew 77.8 percent.[67]

You can now find organic "ready meals," which in food marketing parlance means—yes—organic TV dinners. And you can buy all sorts of organic snacks—organic cheese puffs, organic ketchup, organic Tostitos—and wash them down with organic beer, brewed with organic barley and hops. You can buy nutrition bars and candy. Sales of organic snacks rose by 30 percent from 2002 to 2003. One market consulting firm projected that 120 new organic snack foods would be brought to market in 2005.[68]

Another sign of the growth of organic foods is that almost half of all organic foods are now bought in mainstream supermarkets rather than in specialized "health food" or "natural food" stores.[69] And yet another is that giant food conglomerates have begun colonizing the organic market. The process of mergers and acquisitions is not as far advanced as in the bottled water industry, where three major corporations control over 50 percent of the market, but the trend is unmistakable. General Mills acquired Cascadian Farms, a leading organic brand, in 1999, and it now also owns the Muir Glen brand. Coca-Cola has acquired Odwalla. Smuckers, the jam and jelly giant, owns Santa Cruz Organic. Dean Foods owns Horizon, the organic milk with that cute spotted cow logo, a brand that has over 50 percent of the organic milk market. Say "Heinz," and most people think ketchup, but Heinz now also owns Earth's Best, the organic alternative to Gerber's baby food that I dutifully fed to my three kids when they were babies. Heinz also has a stake in Hain, a maker of organic oils and salad dressings. Hain, in turn, owns Celestial Seasonings, the tea sold in groovy, New Age-y packaging that proclaims the product inside "100% Natural . . . all-natural ingredients and flavors, and no artificial colors or preservatives." Hain Celestial Group's Garden of Eden organic tortilla chips compete with Frito-Lay's organic Tostitos. McDonald's Chipotle Mexican Grill chain offers customers pork from "free-range," naturally raised pigs. Procter & Gamble sells organic fair-trade coffee.[70]

Michael Pollan mentions Dole, ConAgra, and ADM as other food giants that have made the move into organic.[71] When aging New Leftists hear "Dole," they think plantations with underpaid workers toiling under abysmal working conditions, Ugly American companies working hand in glove with the CIA to prop up friendly dictatorial regimes in what were once called banana republics in Central America. ADM and ConAgra are two of the biggest corporations in the grain industry, hardly the kind of companies most people would think of if they ever wondered who actually produces the organic foods they eat. When such big players take notice and begin to move in, it is a clear indicator that a product has become—or is at least well on the way to becoming—a true mass consumer item.

Environmental Activism or Inverted Quarantine Response?

Organic consumption is going up, that much is certain. The question is *why*?

It could be an expression of environmental activism. The consumer chooses organic—is even willing to pay extra to make that choice—because of the following: Growing crops organically means less pesticides and less chemical fertilizers are released into the environment. It means antibiotics are not being given to healthy animals. It means farm animals are treated better. It means strange new bioengineered species are not being planted, with uncertain consequences for other crops or for wild plant species.

It could be, instead, that consumers, made aware of potential hazards to their health, simply want to keep those substances out of their and their kids' bodies.

The apparent cause-and-effect relationship between frightening news and the subsequent jump in demand for organic products seems to suggest the latter. The mainstreaming of organic is often said to have begun with the scare about Alar (pesticide) residue on apples, back in 1989.[72] "An apple a day keeps the doctor away," kids currying favor with their teachers by bringing them a shiny red apple, motherhood and apple pie are all tired clichés, to be sure, but kids *do* eat a lot of apples—fresh apples, apple juice, apple sauce—and parents do tend to believe that it is healthy for their kids to eat fruit. So it came as quite a shock when the public was told that Alar, a known carcinogen, was widely used on apple crops and that residues of Alar were often still on those apples when they were sold and when they were eaten. Similarly, Melanie Dupuis writes that concern about the safety of milk from cows injected with recombinant bovine growth hormone (rBGH) immediately expressed itself as a burst of demand for organic milk. Since then, "sales of organic milk [have] been nothing less than spectacular. . . . the fastest growing organic food segment in the United States."[73] As Pollan writes, "every food scare is followed by a spike in organic sales."[74]

Market surveys bear out this impression that it is mostly the inverted quarantine impulse at work. For a small fraction of all organic consumers, buying organic is a form of political activism; for a much larger fraction, it is not.

The Organic Lifestyle Shopper Study 2000 asked consumers what motivated them to buy organic. Sixty-six percent said they bought organic because it improved their "overall health," 30 percent said specifically because they were concerned about "food safety," and 26 percent said because of the "environment."[75]

The Whole Foods Market survey (2002) asked consumers what "organic" means to them. Seventy-eight percent said it means "products without pesticides," 72 percent said, "products without antibiotics and growth hormones," 68 percent said, "products without GMOs," and 59 percent said, "products without irradiation." One has to go further down the list to get to a social or political meaning, "products grown on a small farm."[76]

Karen Klonsky and Catherine Greene reviewed a small handful of other consumer studies. Hartman Group's 2003 study found that 70 percent of organic consumers say they buy organic because it is healthier. "Protecting the environment did not surface as an important issue in the interviews." Klonsky and Greene's overview of all the survey findings led them to conclude that "health and food safety are the key concerns of consumers."[77]

Michael Pollan, a keen observer of the American agricultural system and of Americans' attitudes toward food, describes two different kinds of organic consumers: "'the true natural' [who is] a committed, activist consumer . . . outwardly directed, socially conscious . . . devoted to the proposition of 'better food for a better planet.' . . . [and] 'health seekers' . . . affluent consumers . . . [who are] more interested in their own health than that of the planet. . . . The chief reason the health seeker will buy organic is for the perceived health benefit."[78]

According to Pollan, "health seekers" already outnumber "true naturals." "[T]rue naturals now represent about 10 percent of the U.S. food market[;] . . . 'health seekers' . . . about a quarter of the market." Pollan also says the size of the first group is stable, while the size of the latter group is growing; thus "the future of organic . . . lies with . . . the health seekers."[79]

Price and Income

Another factor contributing to whether people buy organic foods is their cost. Organic foods typically cost more, sometimes a great deal more, than their conventionally grown counterparts. Some say organic costs more because demand is so strong. Supply lags behind demand, so sellers can charge premium prices.[80] Others argue that organic *has* to cost more because the costs of production are inherently greater: it costs more to use natural pest controls instead of using chemical pesticides; growing crops organically requires more (and more skilled) labor inputs; rotating crops or letting fields go fallow periodically, as organic methods require, reduces the number of profitable crops that can be planted in the same field over any specified period of time.[81]

Polls show that an overwhelming majority of Americans believe that organic foods are more expensive than their conventionally grown equivalents.[82] The very few studies I have been able to find suggest they are right. In 1998, Thompson and Kidwell found that consumers were paying from 40 percent to 175 percent more for various organic fruits and vegetables than they would have to pay for their conventionally grown counterparts.[83] A few years later, Rudy Kortbech-Olesen and Tim Larsen compared the prices of organic versus conventionally grown fruits and vegetables at Whole Foods, Safeway, and one other supermarket. The organic items cost from 11 to 167 percent more than the same foods conventionally grown.[84]

While writing this chapter, I decided to do some price checking myself. I went to Safeway, wrote down the prices of some items I buy regularly: milk, eggs, butter, bread, cooking oil, chicken, hamburger meat, a variety of vegetables and fruit, a box of cereal, and so on. I then went to New Leaf, a small, locally owned chain of organic food stores, and I noted the prices of the least expensive organic versions of the same food items. The prices of some organic fresh fruits and vegetables were roughly similar to the prices of those goods in the mainstream market. For many items, such as rice, potatoes, milk, butter, and eggs, the organic versions cost 25 to 50 percent more than their conventionally grown counterparts. Some items—chicken, steak, and cooking oil—cost twice as much or more. Although my "experiment" would have been more rigorous had I followed the USDA's methodology and priced out the standard "basket of goods" USDA uses to track changes in the "cost of food consumed at home,"[85] the price differentials I saw were similar to the differentials described in the literature, and those differentials confirm the widespread perception that eating organic is expensive.

One would expect, then, to find an income or social class gradient in consumption, and, sure enough, marketing surveys show that gradient quite clearly. People who buy organic foods have on average higher incomes than those who do not.[86] People who buy organic spend more on groceries than those who do not.[87] When asked, about two-thirds of consumers who are not buying organic foods say the most important reason they do not is because it is too expensive,[88] and 86 percent agree that they would "eat organic more often if prices were lower."[89]

* * *

The biomonitoring studies reviewed in the introduction to Part II show that Americans' bodies have in them all kinds of waste chemicals, dozens of pesticides, herbicides, and insecticides, as well as PCBs, dioxins, metals, fire retardants, plasticizers, furans, polycyclic aromatic hydrocarbons, and so on. From the materials reviewed in this chapter, it does seem that at least some of those substances enter our bodies when we eat conventionally grown fruits, vegetables, meat, poultry, and eggs and drink milk from conventionally raised farm animals. People are concerned about the substances they may be ingesting when they eat conventionally grown or raised foods. As their concern has grown, so has their desire to eat, instead, foods that are naturally grown, organic and therefore pesticide free, hormone free, safe.

In this chapter I have again held myself to just describing why people feel at risk and what they have attempted to do about it. In chapters 6 and 7, I take up questions not dealt with here: Does organic food actually "work" as advertised? Is the person who consumes organic foods safer? Do such people have fewer chemicals, pesticide residues, and so on in their bodies? And does this sudden passion for organic food have any significant unintended consequences?

5. Breathing

A person takes in materials from the surrounding environment a third way, breathing. An average-sized adult takes in about 450 to 500 cubic centimeters of air with each breath, from fourteen to twenty times, typically, each minute. Taking the middle of those ranges, that is about 8,000 cubic centimeters of air every minute. In English units that is about three-tenths of a cubic foot of air every minute, or over 400 cubic feet of air per day. If the air has molecules of harmful substances suspended in it, those substances are brought deep into the lung's tiniest passages, either to lodge there or to then enter the bloodstream and be transported throughout the body. Substances can also be absorbed through the skin, and we can think of this other pathway of incorporation as another form of respiration or breathing.

Applying inverted quarantine methods to these sites of exchange with the environment would seem to present a greater challenge than using such methods to control what is taken into the body when eating or drinking because this third kind of incorporation activity is both more pervasive and more diffuse. Food and water are taken in episodically, in distinct, bounded quantities, bites and gulps. Controlling episodic and discrete exchanges at least *appears* to be a more straightforward task. The lungs and skin breathe, engage in exchanges with the environment, continuously. To control food and water, "all" you have to do is make sure each discrete item is clean. Breathing unpolluted air would seem to require control over practically every aspect of the immediate environment.

Is such a daunting task possible? At a minimum, one would have to purchase for one's home a rather large number of alternative, organic, nontoxic, "all natural" consumer items. I was not surprised to learn, once I began to look, that the number of such products has been growing quite rapidly and that there is now a substantial demand for them, though not yet as widespread as the demand for bottled water and organic food. Would buying all those alternative, "natural," and "nontoxic" products do the job?

Here I will follow the same pattern as in the previous chapter, discussing first the risks, then the items people buy to try to shield themselves from those risks.

Exposures

I think it useful to divide exposures via breathing into three categories: outdoor air, indoor air, and products worn on or applied directly to the body.

Outdoor Air Quality

In chapter 3, I wrote that water pollution in the United States is the normal, inevitable consequence of the way we live in our cities, how we grow our food, and how industry produces things that we consume. Water pollution, I wrote, is inscribed in our way of life. Much the same can be said about air pollution. It is not any one thing. Numerous commonplace, everyday activities, acting together, end up degrading outdoor air quality. Consider what the EPA says about the causes of two of our most serious air quality problems, ground-level ozone, better known as smog, and "particulates," fine particles (particles that are ten micrometers or smaller, that is, one-hundredth of a millimeter or smaller) suspended in air:

> Ozone . . . is formed by gases called nitrogen oxides (NO_x) and volatile organic compounds (VOCs) that in the presence of heat and sunlight react to form ozone. . . . NO_x is emitted from motor vehicles, power plants and other sources of combustion. VOCs are emitted from a variety of sources, including motor vehicles, chemical plants, refineries, factories, consumer and commercial products, and other industrial sources.[1]

> Fine particles result from fuel combustion (from motor vehicles, power generation, industrial facilities), residential fireplaces and wood stoves. Fine particles can be formed in the atmosphere from

gases such as sulfur dioxide, nitrogen oxides, and volatile organic compounds. Coarse particles are generally emitted from sources such as vehicles traveling on unpaved roads, materials handling, and crushing and grinding operations, and windblown dust.[2]

Public concern about the health effects of poor air quality, and especially of smog, that pall of yellowish haze over Los Angeles and other large cities, led to passage of the Clean Air Act in 1970. Charged with implementing the act, the EPA put most of its initial effort into setting standards (known as National Ambient Air Quality Standards, or NAAQS) for a handful of so-called criterion pollutants, carbon monoxide (CO), lead, nitrogen dioxide (NO_x), particulate matter, ground-level ozone (O_3) and sulfur dioxide (SO_2), and into measuring compliance with those standards. Areas in compliance are said to be in "attainment."

Since then, a city's or a region's ability to meet standards for these criterion pollutants has been the single most important measure used to assess and discuss air quality. Beyond smog and particulates, beyond the most obvious forms of air pollution, our air also has in it hundreds of other substances of concern. These pollutants are commonly referred to collectively as HAPs, Hazardous Air Pollutants, or as "air toxics." The Clean Air Act, as amended in 1990, required the EPA to regulate 187 different "air toxics," a wide range of substances that include volatile organic chemicals such as benzene, heavy metals, combustion by-products, and solvents.

To reduce HAPs in outdoor air, EPA identifies sources—facilities that emit the HAPs in substantial quantities—and requires these sources to install new control technology (referred to as "maximum available control technology" or MACT) so that discharges of HAPs will decrease over time. A partial list of such sources shows, again, how widespread, how deeply rooted in the very structure of the industrial economy, are the causes of poor outdoor air quality: coal-fired and oil-fired power plants, nonferrous metal (i.e., lead, aluminum, copper, magnesium) processing, ferrous metal processing (coke ovens, steel manufacturing), mineral production, oil and natural gas production and refining, waste treatment and disposal, agricultural chemical production, pharmaceutical production, polymers and resins production.

Smog is the most visible symbol of poor air quality, but dirty air is not just an irritant and an eyesore. It is a serious public health problem.

Since a person takes airborne pollutants into her or his body by breathing them in, concern about the health effects of polluted air focused first on acute and chronic respiratory problems. Research has validated that concern again and again. Breathing polluted air inflames lung tissue, reduces lung function in children, aggravates asthma, increases susceptibility to bronchitis and pneumonia, and increases the risk of getting lung cancer.[3] Subsequently, it has been shown that poor air quality also causes heart disease.[4] HAPs are regulated because they can cause cancer, reproductive and developmental disorders, and neurological problems.[5]

The Clean Air Act has achieved a lot. Nationally, levels of all six criterion pollutants have fallen since 1970; some have fallen quite substantially.[6] Smog is not completely gone. I can testify to that with my own eyes as I drive periodically through San Jose, the big city nearest to where I live. But it is not as thick as it once was, and it reaches unhealthy levels fewer days of the year. Though the available data are somewhat spotty, indications are that levels of nearly two hundred HAPs are also going down.[7]

Still, problems persist. The Government Accountability Office (GAO) reports that the EPA has failed to fully implement the air toxics portions of the 1990 Clean Air Act amendments.[8] Some cities and some regions are still not able to attain the EPA's targets for one or more criterion pollutants. The EPA reports that "as of September 1999, approximately 105 million people [were] living in areas designated as nonattainment for at least one of the criteria pollutants." For 2002, that number was 126 million. The American Lung Association's annual State of the Air reports claim even bigger numbers, in the 150 million range, in 2004, 2005, and 2006.[9]

Even if a city or region is in full attainment and meets all current EPA standards for the criterion pollutants, that does not necessarily mean that the air is completely safe. Those standards may not be stringent enough. Similar to the pattern I described in chapter 3, where I showed how as research improves, regulators find it necessary to keep strengthening drinking water standards, newer epidemiological and clinical studies suggest that levels of smog and particulate matter allowed under current EPA rules are too high.[10]

In fact, having carefully assessed hundreds of new studies, the EPA did propose in 1997 to lower both ozone and particulate matter exposures. The new standards were opposed by powerful industrial associations, such as the National Association of Manufacturers and the American Petroleum Institute. Industry lobbyists claimed "excessive costs, lack of significant demonstrable benefits, loss of competitiveness, and technical infeasibility."[11] In spite of this opposition, the Clinton administration proceeded to adopt the new standards, with some concessions. Industry then challenged the new standards in federal court. The Court of Appeals for the District of Columbia Circuit vacated one of the new standards and remanded others to the EPA for further consideration. The Court of Appeals was subsequently reversed by the Supreme Court, allowing the EPA to implement the new standards.

Health organizations such as the American Lung Association, the American Cancer Society, the American College of Cardiology, the American Academy of Pediatrics, the American Medical Association, and others argue that the evidence supports even stronger particulate standards.[12] More stringent regulations are not likely any time soon. The Bush administration has been proposing rule changes that would weaken enforcement of the Clean Air Act. The administration has proposed, for example, to rewrite the rules governing when industries are required to install the "maximum available control technologies" that reduce emissions of HAPs, and it has proposed to exempt certain rural and farming areas and mining operations from particulate standards.

So, nearly half the population lives in places that have yet to attain compliance with current standards for the criterion pollutants. And there is evidence that air that *is* in compliance is not clean enough; breathing that air can still make people ill. There has been progress since 1970, but outdoor air still is not completely safe to breathe.

Indoor Air Quality

Around 1970, when many of our modern environmental laws were being enacted, neither environmental activists nor policymakers seem to have considered *indoor* air quality an issue. All the talk was about outdoor air—smog, auto exhaust, emissions from factory smokestacks. Reflecting this understanding of the air pollution problem, the Clean Air Act was all about reducing outdoor, ambient air pollution.

Only later did the EPA begin to study indoor air. The results were—or perhaps we ought to say "should have been," since the findings seem to warrant greater attention than these studies have received—eye-opening. The EPA reported that "the air within homes and other buildings can be more seriously polluted than the outdoor air in even the largest and most industrialized cities."[13] Certain contaminants can be present indoors in concentrations five, ten, sometimes a hundred times higher than in outdoor air.[14]

The sources of indoor pollution? Practically everything in the modern home: cleaning supplies, furnishings, the very materials out of which homes are built.

Contemporary standards of cleanliness require regular use of a large array of cleaning agents. We wash our clothes with laundry detergent and sometimes with bleach; we use fabric softeners. We wash dishes with other detergents. We use scouring powders and oven cleaners in the kitchen, scouring powder and toilet bowl cleaner in the bathroom, and liquids or gels that open up clogged drains. Still other products clean wood and linoleum floors, remove stains from carpets, clean windows, polish furniture, and deodorize and "freshen" air. The EPA says that "while people are using products containing organic chemicals [chemicals found in household cleaning products], they can expose themselves and others to very high pollutant levels, and elevated concentrations can persist long in the air long after the activity is completed," and warns that exposure to such chemicals can do "damage to liver, kidney, and central nervous system. Some . . . are suspected or known to cause cancer in humans."[15]

Furniture, such as tables, chairs, bookcases, and bed and sofa frames, especially at the more affordable end of the market, is made with pressed wood products. (Furniture made with whole woods is widely available but is more expensive.) Pressed wood products can "outgas," or "off-gas," formaldehyde long after they have been brought home from the store.[16] Carpets, fabrics that cover sofas and easy chairs, and drapes and curtains are made with synthetic, "man-made" fibers that off-gas when those items are new, and these carpets and fabrics are often treated with chemicals that also off-gas for some time.

The house itself—the very materials out of which the structure is made—can be a source of indoor air pollution. Homes are built with wood products, insulation, adhesives, and paints, all materials that contain hazardous substances. Especially when the home is new or newly remodeled, these materials off-gas.

The names of some of the volatile organic chemicals (VOCs)

being off-gassed are easily recognizable; others less so. Many are known to be hazardous. The VOCs discussed most often in the literature on indoor air quality include formaldehyde (from fabrics, pressed wood, insulation, paper products, carpet backings, adhesives, paints), benzene (glues, paints, stains, solvents), styrene (adhesives, foam, lubricants, plastics in carpet, insulation), para-dichlorobenzene (moth balls, air fresheners, toilet bowl deodorizers), methylene chloride (paint stripper), perchloroethylene (carpet cleaners, stain removers), and carbon tetrachloride (paint remover).

Concentrations of these VOCs can go quite high, especially in homes that are well insulated and during times of the year when windows are kept shut. The EPA also makes the obvious but important point that our lifestyles lead us to spend a lot more time indoors than out.[17]

In 2005, Clean Production Action published Sick of Dust, a study of chemicals found in dust gathered from seventy homes. The researchers looked for—and found—pesticides, phthalates, perfluorinated organics (PFOA and PFOS, used to make Teflon and to make fabrics stain-resistant and water-resistant), polybrominated diphenyl ethers (PBDEs, flame retardant), organotin compounds (found in products made of PVC, polyvinyl chloride, and in polyurethane foam), alkylphenols (in spot remover, detergents, all sorts of cleaning agents, lubricants, etc.).[18]

Products Worn Next to or Applied Directly to the Skin

Moving ever closer to the body, we turn next to clothing and to personal hygiene products. The American Environmental Health Foundation, an online marketer of green, alternative, nontoxic products, asserts that clothing can be hazardous: "The fabrics used in clothing are a particularly insidious source of pollution. Natural fabrics can become contaminated during the manufacturing process when dyes and chemicals are added to them. Synthetic fabrics, which are oil based, are also chemically contaminated."[19]

Nothing gets closer to the skin than personal hygiene products and makeup. Thankfully, women no longer spray a little cloud of vinyl chloride monomer around their heads when they use hairspray, as they did before the use of vinyl chloride as an aerosol propellant was banned in 1974.[20] That ban was an easy call. By 1974 it was generally recognized that vinyl chloride is carcinogenic. Chemical industry officials were more than willing, quite untypically, to see it banned because they realized, as one Union Carbide memo put it, that "a company selling vinyl chloride as an aerosol propellant . . . has essentially unlimited liability to the entire U.S. population."[21] There have not been many other, similar bans. Some of the ingredients in personal hygiene products and makeup, the products we use to clean our skin, wash our hair, make ourselves look good and smell nice, are carcinogens. Some have reproductive effects. Some are known to be hazardous, at least at higher concentrations.[22]

Some of the toxic ingredients found in personal hygiene products are quite familiar to the average citizen. Most folks who took biology in high school will remember formaldehyde, that smelly substance used to preserve the specimens schoolchildren were (probably still are) so loath to dissect. Cosmetics, deodorants, shampoos, and certain hygiene products made of paper (feminine hygiene products, facial tissues) off-gas trace amounts of formaldehyde.[23]

The names of other substances found in personal hygiene products are not as widely recognized by members of the general public, but they are in many products, so exposure to them is widespread. Consider just one, a family of chemicals called phthalates. Phthalates are found in soap, shampoo, deodorant, nail polish and polish remover, facial washes and creams, makeup, and perfume (also in the pacifiers many babies love to suck on). Phthalates are thought to cause cancer and also harm the reproductive system. The biomonitoring studies I cited in the introduction to Part II find measurable levels of phthalates in almost every subject who is tested for them.[24]

Hot Showers

I have already devoted a whole chapter to issues about drinking tap water. It turns out, though, that I was not done with it. Drinking water has another, personal hygiene dimension that I have not discussed but that becomes relevant when we consider breathing through lungs and through skin. We not only drink water from the local utility and cook with it. We also use it for bathing. From childhood, we are taught the importance of bodily cleanliness and are taught to bathe or take showers daily. Besides being good for you, hot showers feel great. Unfortunately, studies show a person who is taking a hot shower is inhaling volatile organic chemicals. VOCs are present because the chlorine that remains in water after it has been through water treatment interacts with organic matter in the water to produce so-called chlorination by-products, including substances known as trihalomethanes.[25] (Chloroform is perhaps the best known of the trihalomethanes.) Before chlorination, municipal water was frequently contaminated with bacteria that cause disease. Outbreaks of waterborne infectious diseases were not uncommon. Chlorination revolutionized water treatment and greatly improved the safety of the public water supply. But, as Joel Tarr points out in *The Search for the Ultimate Sink*,[26] the solution to a troubling problem has sometimes created new problems that were unanticipated and unforeseen when the solution to the earlier problem was discovered, hailed as the answer, and widely and quickly adopted. That seems to have been the case with chlorination and the unforeseen consequence that when we shower, we are inhaling trihalomethanes. Trihalomethanes volatilize quite readily. Think of all that hot water vapor in a shower stall. It feels so great; it is easy to linger a bit, enjoying the sensation, before getting out. Studies show that the body absorbs these VOCs *more easily* when they are either inhaled or are absorbed through the skin than when they are consumed as a drink of water.[27] Trihalomethanes have been linked to bladder cancer, renal cancer, and to a variety of other ailments.[28]

Perception of Risk

Some of the exposures described above are significant. Outdoor air pollutants have been linked to respiratory and cardiovascular problems. And recall the EPA's finding, already cited earlier, that indoor air "can be more seriously polluted than the outdoor air in even the largest and most industrialized cities." Open up a tube of glue, a can of paint stripper or stain remover, and your nose tells you why the manufacturer recommends one use such products only in well-ventilated spaces. Phthalates have serious health effects. They should probably not be found in so many products that we rub on our skins or, say, that we allow babies to suck on.

I do not wish to argue, however, that *every* one of the exposures I have listed here is a significant risk. I am not convinced, for example, that there is really that much to fear from clothes made of synthetic fabrics. Yes, certain dyes and the chemicals used to make fabrics wrinkle-free—"permanent press"—do off-gas formaldehyde. Dry-cleaned clothes can off-gas perchloroethylene (perc), the carcinogenic VOC used by dry cleaners (appropriate, here, to have an olfactory memory of the smell when one walks into the dry cleaners to pick up one's clothes). But I am not prepared to accept that the clothes we wear are really that much of a threat, or that it is reasonable for the American Environmental Health Foundation to use the ominous phrase "insidious source of pollution" when writing about conventional clothing.

The risks associated with some of these exposures may be trivial and contribute little or nothing to the total risk associated with *all* the toxic hazards to which a person is regularly exposed. I am inclined to believe that some of them may be almost entirely imaginary. But the question is not what *I* think. What matters is that lots of people—an increasing number of people—consider these risks real, real enough that they are prepared to spend the time and the money to buy products that promise to reduce their exposures.

Commodities

"Natural" furniture, "natural" clothing, "green" household cleaning products, "natural" makeup—the market is flooded with products that promise explicitly that if a person buys them and uses them, he or she will be exposed to far lower amounts of toxic substances and therefore will absorb through lungs and skin far fewer molecules of those substances. Such products are no longer just for oddballs who are maybe a bit too preoccupied with their health or just for persons who suffer from MCS, Multiple Chemical Sensitivity. Like bottled water and organic food, though not yet as widespread (perhaps only lagging in time somewhat), many of these items are going mainstream.

Outdoor Air

Actually, short of purchasing the kind of protective gear donned by chemical workers or people hired to clean up hazardous materials spills, there is not much that a person can buy, even today, to protect oneself from outdoor air pollution. The typical air filter one can buy for one's home or for one's car is not designed to remove toxic chemical pollutants coming into the house or the car from outside. Reading the advertisements, I see that these air filters claim only to remove bacteria, viruses, mold spores, pollen, dust mite allergens, animal dander, and second-hand cigarette smoke.[29]

If a person wishes to breathe cleaner outdoor air—as a consuming individual, I mean, not as a political actor who, say, pays dues to an organization that lobbies to strengthen the Clean Air Act or votes for candidates who support clean air initiatives—the answer is to live in a place that has better air than other places.

One can interpret the findings of what is known as "environmental justice" research as evidence that residential location has an environmental inverted quarantine dimension (in addition to the *social* inverted quarantine dimension discussed in chapter 2). Almost universally, one finds disproportionately larger numbers of the poor, poorly paid workers, and people of color living closer to hazardous waste facilities and polluting industries.[30] Although environmental justice research rarely pays much attention to the other end of the class hierarchy, the data show that the opposite is also true: the farther a census tract is from toxic waste sites or from the local industrial district, therefore where the air is presumably cleaner, the more affluent (and more white) the residents of that tract.

Many things, of course, go into the decision about where one lives, for example, proximity to job and family, proximity to shopping, entertainment, and cultural amenities, and the reputation of neighborhood schools. The data clearly show, however, that when people pay more for a home in a residential neighborhood, they are also paying for the benefit of not having to live near toxic industrial facilities (dubbed in the literature LULUs, for Locally Unwanted Land Uses). Here the inverted quarantine commodity purchased is not the house per se but the *location*. The anticipated protection is conferred not by a physical barrier, such as an actual filter, or by a substitute commodity that claims to be uncontaminated. The barrier, here, is simply distance, spatial separation. Since the commodities involved, land and house, are sold and bought on the real estate market, social class position inevitably is fundamental to one's ability to indulge in inverted quarantine at this level.

Indoor Air

Exercising the inverted quarantine option in order to secure for oneself better outdoor air requires one to make one of the biggest decisions a person is ever asked to make: Where should I live? Where should I buy a house? The inverted quarantine approach to indoor air pollution requires of one far less momentous choices, though still potentially expensive ones. One only need go out and buy a number of consumer items that, unlike their conventional counterparts, will not off-gas hazardous VOCs in one's house. Books advise consumers about benign alternatives to using chemical detergents, stain removers, oven and toilet bowl cleaners, and so on.[31] Companies with names like Ecover, Naturally Yours, Seventh Generation, Earth Power, Earth Friendly, and Bioshield sell "natural" or "nontoxic" alternatives for every conceivable kind of household cleaning product. It is easy to find these and other, similar products on the Web.[32]

If a person is at all aware of the world of green alternative products, they will have heard about nontoxic household cleaning products. Those products are, however, just the tip of the iceberg. Want to stay safe while you do home improvements? You can buy nontoxic adhesives, sealers, primer, paints, and stain. If you can afford it, organic home furnishings are available too: bookshelves, desks, dressers, computer tables, and home office furniture made of real wood, not wood fiber or particle board—all wood, no glue. Go to www.nontoxic.com, and you can order "organic furniture for sitting room or library . . . pure wool carpets . . . organic mattresses . . . bedding . . . wool comforters and pillows, futons, natural mattress pads."

Products Next to the Body

One can find organic clothing in specialty stores. Kids Nature, a few blocks from my house, sells "the healthiest clothes available today . . . made out of organic cotton, untreated wool or linen, . . . contains no pesticides or other harmful chemicals like conventional clothing."[33] One does not need to live in a hip Northern California community to have ready access to such goods. As with so many things, the Web has made them available to consumers anywhere. In 2003, when I wrote the first draft of this chapter, a Google search for "organic clothing" returned over 3,000 hits. I checked again in February 2005; "organic clothing" returned 42,900 hits. By June 2006, that number had risen to 239,000. Some of that increase is surely due to improvements in Google's search algorithm, but it cannot be only due to that; some of it must reflect real growth.

There are "organic" alternatives to almost every mainstream personal hygiene product. I browse through a recent issue of "Taste for Life," a glossy publication devoted to healthy, organic living, distributed free by one of the organic food stores in my town. In one issue I find ads for organic sunscreens, shampoos, and conditioners and for "Kiss My Face" skin products, "obsessively organic face care" that promises to "feed your skin the way NATURE intended."[34] Go to the American Environmental Health Foundation Web site, and a few clicks of the mouse will take you to organic personal hygiene products of every kind: "Deodorants . . . Bath/Shower Moisturizers . . . Eye Care . . . Face Care . . . Hair Care . . . Hand and Body Cleaners . . . Mouth and Tooth Care . . . Powder . . . Sunscreen."[35] The Web site www.shopnatural.com claims to offer five thousand different organic products. The VOCs-in-the-hot-steamy-shower problem is eliminated by installing a water filter above the shower head. These filters are sold by the same companies that market under-the-sink water filter units.

Market data for these products, data on growth trends, for example, are somewhat spotty. The handful of references I have been able to locate suggest healthy growth rates, on the order of 20 percent or more a year, and a pattern of mainstreaming—"green" and "nontoxic" products increasingly found on the shelves in conventional, mainstream supermarkets—similar to what we see with organic food.[36]

Web-based marketing must certainly facilitate growth. A decade ago, one had access to such products only in special stores or via the mail, purchased from a handful of catalogs such as the Seventh Generation and Real Goods. Today, anyone with a PC, a modem, and a credit card is only a few clicks away from finding, learning about, and purchasing these goods.

Income, or Social Class

Finally, as always, disposable income is the key to who can make these consumer choices and who cannot. You can reduce exposure to outdoor air pollution only by spending hundreds of thousands of dollars more for a home away from the industrial zone and the traffic corridor. In the home, every organic alternative costs more—and sometimes a great deal more—than its "conventional" counterpart. The organic mattresses offered at www.nontoxic.com sell for around $1,000, far more than a person pays for a mattress at, say, Macy's or at a discount mattress and futon store. Clothing at Kids Nature costs more than kids' clothing at Kmart, Target, or JCPenney. Nontoxic cleaning agents cost more. Organic makeup costs more. And it is not just the extra cost of any one product.

If one believes that one should buy self-protection from hazardous chemicals that can enter the body via breathing or by being absorbed through the skin, one has to buy *lots* of products, *all* the things I have listed above. A middle-class person may purchase one, two, or a few of these items: organic soap and shampoo, a nontoxic detergent, an alternative to the usual oven cleaners. It is better than nothing, but if one buys only one or two such items, that will not help with all the *other* potential sources of contamination in the home. The closer a person can get to filling their home with every conceivable natural, organic, or nontoxic alternative, the higher the apparent likelihood of protection. But when every alternative product costs more, it takes ever higher levels of disposable income to keep making that choice again and again. As with bottled water and organic food, here, too, a class gradient is the inevitable corollary of a consumer orientation to risk.

Some people, but not the vast majority, have that kind of money. They are able to buy $1,000 mattresses, organic bedding, natural wood furniture, rugs, nontoxic household supplies, organic makeup—everything. For them, a special market segment has now emerged. The marketing world calls this new market niche, offering products for folks who want to conspicuously consume organic, "organic style." Rodale Press, publisher of the well-known magazine *Organic Gardening,* has a new publication, *Organic Style,* for "conscious sensualists," wealthy women who, according to Maria Rodale,

> want to do the right thing for their health and the environment, but not at the cost of living well. . . . They do not want to sacrifice anything. Not great food, great clothes, nor a comfortable home that looks good.[37]

Well, who would not want to do that? "Do the right thing" for the environment while one continues to live the affluent lifestyle, with its high levels of consumption? Not that many people can afford to have it all.

* * *

Again, let's recall that the biomonitoring studies I reviewed in the introduction to Part II show that Americans' bodies have in them traces of all sorts of household products, phthalates, disinfectants, aromatic hydrocarbons, the perfluorinated chemicals found in

Scotchgard and used to make Teflon, and so on. Many of these substances get into our bodies as we breathe air outdoors and indoors, and also through our skins as we come in contact with products that contain them. Even before the biomonitoring studies confirmed it, Americans suspected that they were breathing in toxics and absorbing toxics through touch, and those who could afford it began to try to bring more natural and chemical-free products, ranging from furniture to cleaning supplies and personal hygiene products, into their homes and bathrooms.

As in the previous two chapters, I have restricted myself here to simply describing the risks and the products that promise to shield consumers from those risks, and have refrained from more analytic and critical discussions of efficacy and of possible unintended consequences. I take up those questions next, in the final part of the book.

III
Consequences of
Inverted Quarantine

Consequences of
Inverted Quarantine

In Part III, I turn from description to analysis, to questions of effectiveness, consequences, and prospects. Are environmental inverted quarantine products effective? What will happen if millions of people continue to seek to protect themselves by using these products?

In chapter 6, I consider these products' effectiveness: do they work as promised and as hoped? The evidence, though scant, suggests not. Yes, some of these products provide an increment of protection, but no amount of vigilance can come even close to completely, perfectly keeping from the body all the hazardous substances now circulating in our environment.

Do these products work well? That is a question that matters greatly to those who buy these products, hoping that that will protect their health and their children's health. For me, there is a more important question: What happens to society in the long run when millions of citizens *believe* (rightly or wrongly) they can protect themselves individually from environmental hazards? That is the question that goes to the heart of my concern about the inverted quarantine phenomenon. I try to answer it in chapter 7.

Since such products have only been on the market a short time, historically speaking, any discussion of long-term consequences must necessarily be, at this point, speculative. Still, we know quite a lot about the environmental conditions that exist today, and scientists are able to make informed predictions about how those conditions are likely to evolve over time. We know how people tend to respond to perceived risk and about how our political system works. We have the examples of the fallout shelter and of suburbanization, exurban sprawl, and gated communities. Both, as we have seen, are rich sources of insight into both the societal impacts and the longer-term effectiveness of inverted quarantine strategies.

In chapter 7 I bring all of these diverse pieces of knowledge together in an effort to discern likely long-term consequences. I conclude that mass flight into inverted quarantine decreases the likelihood—and defers the day—that something substantive is done about those hazards. Robbed of a potential mass constituency of highly motivated supporters, environmental politics languishes and conditions worsen.

Chapter 7 argues that this is already happening and will continue to happen, with increasingly troubling implications, if people continue to retreat into little consumer bunkers, thinking they can shield themselves from environmental harm by consuming only pure, organic, "natural" goods. But will they? Or will Americans eventually reject inverted quarantine, as they did at the end of the fallout shelter panic of 1961? Those are the questions I take up in the conclusion that follows Part III. The fact that people have been willing to reject inverted quarantine before, always with excellent results, is certainly cause for optimism. On the other hand, an assessment of dominant tendencies in our political culture—how Americans are reacting to all sorts of threats today—leads me to conclude, reluctantly, that in the immediate future we are likely to see more—and increasingly desperate—acts of inverted quarantine.

6. Imaginary Refuge

All this green and natural consuming, all this filtering—does it work? Can it work? If people use these products, do they really manage to keep toxic substances out of their bodies?

I begin by looking at what is known about the effectiveness of individual inverted quarantine products. I find that their effectiveness varies from "substantial" to "apparently none," and in many instances their effectiveness cannot be determined at present. Of course, since the body takes in materials from the environment a number of ways, no single product, no matter how good, is enough by itself. Most consumers are not implementing the full program, however. They are not buying and using *all* the inverted quarantine products available, all the time. For one thing, it is just too expensive. Most people cannot afford it. The inevitable conclusion, given the two facts that individual products vary widely in effectiveness and that most consumers do not and cannot buy them all, is that the average user is currently getting at best only a modest increment of protection.

I explore what it would take to make the inverted quarantine approach actually work. A person bent on shielding themselves would have to buy and use many such products, and the products would have to be much better than they are today. If all the individual products *were* that good—and they are not—a person would have to commit an immense amount of time, attention, and money to the effort. As the search for protection intensified, the monetary and psychic costs would increase exponentially. Even then, I will argue, there are absolute limits, and it is not possible to achieve, or even approach, complete protection.

Assessing Inverted Quarantine Products' Effectiveness

Bottled Water

Bottled water was the first inverted quarantine product to achieve mass consumer item status. Of all the products I have described, it is the one that has been studied most extensively. The findings, though not particularly shocking, are not that reassuring either. Researchers find bacteria, organic chemicals, and inorganic chemicals in samples of the bottled waters they test. For the most part, these contaminants are present in small amounts, so the product complies with applicable state and federal drinking water standards. (But recall from chapter 3 that if standards are inadequate, being in compliance with them does not mean that that water is clean and healthy. It only means that the content of that bottle is no worse than tap water, not that it is better.) Researchers have also consistently found that a small but significant percentage of the bottled waters they have tested actually exceed one or more drinking water standard. Bottled water may taste better than tap water, but in terms of chemical or biological content it is not obviously superior to tap water, and sometimes it is demonstrably worse.

Problems were evident even in the earliest studies of bottled water quality, done in California and New York, two states whose residents started buying bottled water in quantity some years before people elsewhere in America took to it. In California, researchers found significant amounts of inorganic chemicals, such as arsenic, and organic chemicals, such as benzene, phenols, and chloroform, in the samples they tested. A review of plant inspection records revealed all sorts of problems, not just run-of-the-mill quality-control problems, but apparently outright fraud:

> not monitoring the quality of their water sources . . . unsanitary maintenance of facilities, lack of personnel training in sanitary methods of bottled water production . . . not keeping required records of water testing results, and for keeping false records which indicate more monitoring than is actually performed . . . doctoring water samples with chlorine so that laboratory results would show no bacterial growth . . . adulteration, misbranding and misrepresentation of their product.[1]

The report concluded that "under current quality standards, bottled water or vended water is not guaranteed to be a safe and wholesome product."[2]

The New York studies also reported problems. More than half of the bottles tested by the state's Department of Health "contain volatile organic chemical contaminants but at concentrations below current drinking water standards or guidelines."[3] The Suffolk County, New York, Department of Health Services found, similarly, that

> most bottled waters . . . meet current drinking water standards. However, . . . significant levels of synthetic organics, primarily trihalomethanes, 1,1,1 trichloroethane and toluene were present in almost 50 percent of the brands tested.[4]

In a follow-up study, Suffolk County officials reported that seven of the seventy-nine brands tested—almost one in ten—"exceeded a drinking water standard." The report went on to observe that "this is not an impressive number for an industry from which the public generally expects a product of higher quality," and concluded that in some cases "it would appear that tap water from a public supply may be the better choice."[5]

These were just state and county health department reports. They were not widely disseminated nor widely publicized. Bottled water sales boomed.

Then, in 1990, benzene, a carcinogen, was found in bottles of Perrier. Perrier's troubles were big news. This was *Perrier*, the industry's star brand, sponsor of glitzy sports superevents like the New York Marathon and the 1984 Olympic Games in Los Angeles.[6] An article in the *New York Times* Business Section recalled that back then Perrier was the

> status symbol . . . [for] the trendy yuppie of the early 1980s . . . The sparkling mineral water from France was a must-have social drink, quaffed by celebrities like the tennis star Chris Evert and the fashion designer Halston and served at the best clubs, from the wild Studio 54 to the chic Regine's.[7]

Two hundred eighty million bottles of Perrier were recalled.[8] The problem was soon fixed, but the damage was done. Sales of Perrier fell by half.[9] Perrier's misfortune threatened the whole industry. In an article titled "Consumer Confidence: Your Most Important Ingredient," a beverage industry market publication asked, "Has the sanctity of bottled water been called into question?"[10] Well, yes, it had. Bottled water's "sanctity" rested on the belief that this was pure water, nothing but water, not like that stuff that flows from the tap. But now the industry's marquee brand was found to have been contaminated with a known carcinogen.

Benzene had been in bottles of Perrier water for over eight months, undetected. The 1990 Suffolk County report on bottled water quality pointed out that "the greater frequency of monitoring for volatile organic compounds required of most community water supplies would make it extremely unlikely that a significant benzene contamination of a public water supply would go undetected for a six month period."[11]

The "sanctity" of bottled water had, indeed, been called into question. Here was a widely publicized instance where tap water turned out to be far safer than the most famous brand of bottled water. The consumption figures tell the story. Bottled water consumption had been growing vigorously every year; in 1990, sales suddenly went flat.[12]

Benzene in Perrier was a real scandal, the kind of event that makes Congress spring into action. The House Commerce committee decided it was time to investigate the industry.[13] The committee learned that federal Food and Drug Administration officials had been doing a poor job regulating the industry.

In relation to standards, the committee found:

FDA has not adopted [for bottled water] . . . all the health-based public drinking water standards established by the EPA that set maximum levels for certain harmful contaminants.

As a result, bottled water . . . may contain levels of potentially harmful contaminants that are not allowed in public drinking water.[14]

In relation to inspections, the committee found:

The agency . . . does not consider bottled water to be a priority, and therefore inspects bottling plants infrequently. . . . on average, the FDA inspectors visit plants every 5.75 years.[15]

FDA does not have a compete inventory of bottlers; it may not have inspected some domestic plants. . . . Further, [when FDA does inspect a plant] of the 31 contaminants for which there are standards, FDA tested for 5 or fewer contaminants in 94 percent of the tests.[16]

Committee Chair John Dingell (D-MI) declared the FDA's performance "deplorable" and "inexcusably negligent."[17]

At the state level, only two states, New York and California, seemed to have reasonably solid regulatory programs in place.[18] A handful of other states had adopted the International Bottled Water Association's (IBWA's) Model Code, but they did not budget for enough inspectors, so bottling plants were rarely visited.[19] (A decade later, the Natural Resources Defense Council reported that forty-three states had less than one staff person assigned to monitor compliance with the state's bottled water regulations.)[20]

What about industry self-regulation? The IBWA's CEO testified that the industry had

> a good story to tell . . . The industry has been very aggressive about its policy of self-regulation and adherence to stringent purity standards. . . . The philosophy of our members is to know and control exactly what goes in the container.[21]

The congressmen were skeptical. When Perrier water was contaminated with benzene, the problem was not detected by the company but by a state laboratory.[22] Chairman Dingell pointed out that that was typical of other recalls:

> In 1990 alone, there were 22 bottled water products recalled . . . [and] It is worth noting that these contaminants were not detected by the bottlers themselves, but by consumer complaints and by State laboratories. . . . The industry [claims] that its self-regulation is unparalleled in the food industry . . . the subcommittee has concerns about the rigor and the effectiveness of IBWA's self-policing.[23]

The hearings showed that bottled water products were subject to far weaker regulation than ordinary tap water was. Tap water had to be in compliance with a larger number of health standards. Tap water was tested much more frequently and more rigorously.[24]

Bottled water quality could be compromised at any stage in the production process. The source water itself could be contaminated with minerals that had leached from surrounding rock formations, with chemicals from farming, industry, and so on, or with bacteria. Contaminants could be introduced while the water was being processed and bottled.[25]

Thus, if regulation was weak, products of questionable quality could reach the market.[26] Several reports were submitted for a workshop organized by the subcommittee. They reported that some bottled water products had in them inorganic chemicals (aluminum, cadmium, chloride, magnesium, mercury, etc.), organic chemicals, and/or high levels of bacteria.[27] Reviewing the literature on bottled water quality, a spokesperson for Friends of the Earth summed up what the research showed:

> Actual exceedences of drinking water standards may be infrequent overall. The extent to which contaminants are found at lower levels, on the other hand, appears to be much greater. . . . bottled water frequently contains low levels of contaminants such as heavy metals and organic solvents.[28]

The spokesperson stated in conclusion that if consumers think that "bottled water is invariably preferable to tap water[, f]rom a health perspective, this is an unfounded assumption . . . despite the attractive packaging of bottled water, this product is not necessarily any safer than the water which comes out of most faucets."[29]

Regulation did improve some after the 1991 hearings. As the EPA updated its standards for tap water, the FDA made more of an effort to keep pace and require bottlers to meet those newer, more stringent standards. The FDA issued rules that defined what terms like *spring* and *mineral* meant, thereby bringing more consistency and honesty to product labeling. More states adopted IBWA's Model Code (but, as already noted, they then failed to adequately fund enforcement of these regulations).

If those modest improvements in regulating bottled waters had done any good, it should have shown up in the data. In 1999 the Natural Resources Defense Council (NRDC) published the most detailed and extensive study of bottled water quality I know of. NRDC paid independent laboratories to test one thousand bottles of 103 different brands of bottled water. The study found that

> Nearly one in four of the waters tested (23 of the 103 waters, or 22 percent) violated strict applicable state (California) limits for bottled water in at least one sample, most commonly for arsenic or certain cancer-causing man-made ("synthetic") organic compounds. . . . Nearly one in five tested waters (18 of the 103, or 17 percent) contained, in at least one sample, more bacteria than allowed under microbiological-purity "guidelines" (unenforceable sanitation guidelines based on heterotrophic-plate-count [HPC] bacteria levels in the water) adopted by some states, the industry, and the EU [European Union]. . . . In sum, approximately one third of the tested waters (34 of 103 waters, or 33 percent) violated

an enforceable state standard or exceeded microbiological-purity guidelines, or both, in at least one sample.[30]

Over several decades of research, then, the findings have been remarkably consistent. Many samples of bottled water have detectable amounts of chemicals and/or bacteria in them. Mostly, these harmful substances are present at low levels, and the product is in compliance with applicable standards. Some samples have levels of contaminants that exceed those standards.

The problem is not with one or two brands that are consistently subpar. If that were so, the solution would be easy: tell consumers to avoid those brands, or just close down the bad-apple bottling plants. The problem is more complex and more intractable. In several studies, the brands that had a problem on the initial tests turned up clean when the tests were repeated on different samples of the same brand. This suggests that the problem is not one or two rogue brands but that the industry suffers from quality-control problems.

Can the average consumer be confident that the next bottle she or he pulls off the shelf will in fact contain a pure product? It is likely to be OK, but it may not be. The water in *this* next bottle may have a high bacterial count, or it may have unacceptably high levels of some hazardous chemical. The consumer cannot test each container. Each bottle demands of the consumer a small leap of faith. The data suggest, though, that that faith might not always be warranted. What is in that bottle is probably about as good as tap water, maybe better, sometimes worse.

Filtered Water

Home filters use a variety of technologies to improve tap water quality. Each technology removes some kinds of contaminants and fails to remove others. Carbon filters, the most common type of home filter technology, remove (actually, reduce the concentrations of but do not completely remove) certain bacteria, organic contaminants such as chlorine and VOCs, and lead but not inorganic contaminants such as salts and metals. Reverse osmosis can remove (reduce) inorganic contaminants (salts, metals), dissolved solids, nitrates, bacteria, and viruses. Distillation removes salts, metals, and minerals but can actually increase the concentrations of organic chemicals.[31]

That is *in theory*. The GAO studied water filters in 1991 and re-ported troubling facts about the state of the industry. Home water filter units were not regulated by the federal government. Only three states, California, Iowa, and Wisconsin, regulated these products.[32] NSF International, a respected independent testing or-ganization, set standards for all the different water filter technolo-gies and certified products to be in compliance with those stan-dards when firms voluntarily submitted their products for testing. The GAO found that "only 43 of the estimated 600 manufacturers had had units certified" by NSF. "While some of these companies are among the industry's largest," the GAO continued, "most cer-tified units are designed only to improve the aesthetic properties of drinking water." Those facts suggest that consumers were buying a product whose health claims were, at best, unproven. The GAO also reported that unfair and deceptive marketing practices, "use [of] scare tactics . . . overstat[ing] the capacity of their units . . . falsely claim[ing] their products are approved by the government," were common.[33]

Well, if filtered water is not always better than tap water, at least it is not worse, right? Maybe not. If the carbon filter is not changed in time and becomes saturated, it can behave like a "high-tech equivalent of a dirty sponge,"[34] and begin to release the contami-nants it had filtered out, now in doses much higher than when the water first came into the house.[35] The GAO also wrote that some units, poorly built, "may leach harmful contaminants into drink-ing water from materials used to construct the units."[36]

Water filter performance may have improved since. Although the federal government still does not regulate water filter products, and only two more states, Colorado and Massachusetts, have joined the three that had regulations in 1991,[37] NSF says that all or nearly all of the really big firms in the industry have had their products certified by it.[38]

We should remember, though, that each water filter technology deals with only some kinds of contaminants. Few homes run their water through a carbon filter, first, then through reverse osmosis, and so on. Second, I note that NSF requires products to *reduce* contaminants to some specified level, not to *eliminate* those con-taminants,[39] so certification means a filter merely reduces expo-sure; it will not eliminate exposure.

NSF itself says, on a Web site titled, "Drinking Water Myths," that it is a "myth" that "using a home water treatment device will make tap water safer or healthier to drink." The "reality" is that "many of the water treatment devices . . . can help improve the taste, odor, or color of the water. . . . some of these products can actually reduce our exposure to harmful contaminants such as lead, parasites *(Cryptosporidium)*, and trihalomethanes (chlorination by-products)."[40]

In sum, water filters can eliminate some contaminants from tap water, but the water that flows from them is not completely pure. That is when the certified filter device is properly used. Water from a carbon filter, the water filter technology most homes have, can actually be more contaminated than the tap water it is supposedly purifying if the consumer does not keep good track of the when they last changed the filter (and it is not that hard to imagine that some homeowners, busy and harried, do lose track of when they last did that).

Organic Food

I have been able to locate only a few—surprisingly few—studies that empirically test the purported benefits of an organic diet. The findings, though somewhat thin, are encouraging.

One important claim seems to be holding up well: consuming organically grown food does significantly lower one's exposure to pesticide residues. One study found that conventionally grown produce is three times more likely than organically grown produce to have pesticide residues on it, 73 percent of samples as compared with 23 percent. Samples of conventionally grown produce were six times more likely than organically grown produce to have *multiple* residues on them.[41] A second study went further and actually looked for pesticide by-products not on the produce but in consumers' (in this case children's) bodies. Kids who ate mostly organic had one-sixth the level of pesticide by-products in their bodies than did kids who ate conventional foods.[42] Eating organic fruits and vegetables seems to offer real, substantial, though not complete, protection.

The findings pose an interesting question. If no pesticides are sprayed on organically grown plants, where do the small amounts of pesticide residues found on or in them come from? They can get there through the air, blown in from fields where pesticides are applied to conventionally grown crops.[43] (Similarly, pollen from genetically modified crops can drift downwind, so it sometimes turns out that organically grown crops become, contrary to the intentions of the organic farmer, genetically modified crops.)[44] Pesticides can also get there via irrigation water, which is often contaminated, as we saw in chapter 4. Pesticides can even arrive via rainwater: the U.S. Geological Survey reports finding pesticides in the rain that falls in California's Central Valley.[45]

Although I have found no studies that definitively prove it, I believe eating organic meat, chicken, and eggs reduces a person's exposure to hormones and pesticides. The animals are not injected with hormones. They get organic feed, so their bodies are likely to have fewer pesticide residues in them (though still some, considering the studies just cited, that found that organically grown food plants do still have some pesticide residues on them).

Other Products

The federal government sets standards for drinking water, so claims made for the superiority of bottled water can be tested empirically. The USDA regulations specify how crops must be grown and how farm animals must be raised if they are to be marketed as "organic."[46] What one gets when one buys alternative, "eco-friendly," or "natural" household cleaning products or alternative hygiene products (soap, shampoos, sunscreens) and cosmetics is less clear. Some of those products might be made of less toxic materials and might be safer and healthier to use; others might not. The average consumer may have trouble determining if it is one or the other.

A popular cleaning product claims to be safer than other, similar products on the market. Some checking by a journalist reveals that its "key ingredient [butyl cellosolve] is the same toxic solvent that can be found in traditional all-purpose cleaners such as Formula 409 and Windex."[47] How common is that?

I would like to believe that if I use an ecofriendly cleaning product, the air in my home will be less toxic than if I had used that product's more conventional counterpart. I would like to believe that if I wash my kids' clothes with laundry soap that proclaims itself "free" and "clear," the air in my home will be cleaner because no VOCs were volatilized by the hot water in the washing machine, and that the earth will be better off because somewhat less detergent will be flushed into the wastewater system.

For some products, the claims seem to be valid. AlterNet and

other online sources of information on alternative cleaning products list materials that one can either make oneself or buy from
companies such as Ecover and Seventh Generation that, the advice
sites say, *are* really less toxic and equally effective. The AlterNet
site also warns,

> The word "natural" is undefined and unregulated by the govern-
> ment and can be applied to just about anything under the sun—
> including plastic, which comes from naturally occurring petroleum.
> Because no standards exist, claims such as "non-toxic," "eco-safe,"
> and "environmentally friendly" are also meaningless.[48]

The Consumer Union's online "Guide to Environmental Labels"
has "report cards" that let consumers know if a particular eco-
claim on a product label is regulated or not, and whether an independent organization is behind that labeling claim or not. In
many instances, the answers are "no" and "no."[49] Consumer Union
tells its readers that the Federal Trade Commission issues "Guides
for the Use of Environmental Marketing Claims (Green Guides) to
reduce the misuse of these environmental labels in marketing and
confusion for consumers[,] but there is no verification system for
most of these labels."[50]

The situation for personal hygiene products and cosmetics is hardly
better. Consider the case of "natural" cosmetics. Here are some ex-
cerpts from an FDA Web site that discusses the safety of cosmetics:

> "Natural" can mean anything to anybody. . . . "There are no stan-
> dards for what natural means," says [John E.] Bailey [director of
> FDA's Office of Colors and Cosmetics], "They could wave a tube
> [of plant extract] over the bottle and declare it natural. Who's to
> say what they're actually using?" . . . Revlon, Inc., uses natural
> plant extracts in its New Age Naturals cosmetics line, says Dan
> Moriarity, Revlon's director of public relations. "But the base for-
> mulas are the same as our conventional products," he says. . . .
> In addition, natural does not mean pure or clean or perfect ei-
> ther. According to the cosmetic trade journal Drug and Cosmetic
> Industry, "all plants [including those used in cosmetics] can be
> heavily contaminated with bacteria, and pesticides and chemical
> fertilizers are widely used to improve crop yields."[51]

In 2005, in response to a lawsuit brought by the Organic Consumers Association and some individual firms that make and sell

organic personal hygiene products, the USDA agreed to certify personal hygiene products as "organic" if their ingredients were 95 percent or higher organic. The association and those individual firms sued because, as an employee of the association said, "a number of industries [are] making millions by making misleading claims."[52] The plaintiffs believed that the USDA "organic" label would offer consumers a way to distinguish products that are actually made of organic ingredients from products that are made with conventional ingredients but have nice-sounding but essentially meaningless claims on their labels.

Access to the USDA "organic" designation was a step forward. However, it does not fully guarantee that consumers, who buy such products because they do not want to breathe in or absorb through their skins potentially harmful chemical substances, get the kind of protection they may think they are getting:

> "It's really very hard to make a shampoo or a skin-care product that is 95 percent organic," said Morris Shriftman, the senior vice president of Avalon Organics. "There are ingredients in those products that are not organic, and those are the things that clean your skin or get under the grime that is in your hair." In shampoos and soaps, rich lather, clean scent and long shelf life often come courtesy of synthetic surface-active substances, perfumes and preservatives. Moisturizers . . . often contain petrolatum, a gelatinous substance derived from petroleum, and emollients like dimethicone.[53]

For that reason, makers of personal hygiene products are more likely to apply for the USDA's "made with organic" certification, which requires only that 70 percent of the ingredients be organic. A consumer has to be paying close attention to catch the difference between a label that says "organic" and one that says "made with organic ingredients." To muddy the picture even more, nothing in the new certification process keeps a product from calling itself "organic," completely regardless of its actual contents, if the word "organic" is part of its trademarked name.[54]

A product may comply with standards set by a reputable international testing agency and still not actually give the kind of protection a person thinks they are getting when they buy that product. In chapter 5, for example, we learned that the domestic water supply has chlorination by-products—volatile organic chemicals, trihalomethanes such as chloroform—in it when it comes into the home. As the name implies, VOCs volatilize readily. VOCs are present in the steam when we shower, and they are absorbed by lungs and skin. We know trihalomethanes have been linked to a number of serious illnesses. Does installing a water filter above the showerhead help? Checking the NSF Web site, I found that in 2005 NSF issued a standard for shower filters. It sets a standard, however, only for a filter's ability to remove "free available chlorine," not chlorinated by-products. It is, as NSF says, an aesthetic standard, not a health standard.[55] A showerhead filter may remove trihalomethanes from the steam, but it qualifies for NSF certification even if it does not.

All that said, I do believe that some of these alternative products can reduce a person's exposure, even if we don't have the data to prove it. If a person buys furniture made of whole wood, as opposed to "wood product" (that is, made of wood chips and fiber held together with some kind of glue), rugs made of natural fiber, and so forth, the air in their home must have lower levels of formaldehyde or chloroform than homes that have furniture made of particleboard and rugs made of synthetic fibers.

Based on this survey of what we know—and do not know—about these products, I conclude that their ability to shield a person from environmental hazards varies greatly. Bottled water, my personal icon/prototype of the environmental inverted quarantine commodity, has been shown again and again to be of questionable value. The inverted quarantine product that has been studied most thoroughly proves unreliable; that tends to cast a shadow on the whole enterprise. On the other hand, organic food seems to work as promised, and in fact works better and better as one increasingly substitutes organic for conventionally grown foods in one's diet. The rest vary from promising to doubtful, with considerable uncertainty; often there is no way to be sure, really, what good they actually do.

What Would It Take for Inverted Quarantine to Work?

The facts I assembled in chapters 3 through 5 will not surprise anyone who has spent time informing themselves about everyday life in modern times. It is unfortunate, but undeniable, that we eat, drink, and breathe toxic materials, in tiny amounts, all the time. In these circumstances, what would it take for the inverted quarantine approach to work?

First, contrary to the evidence I have just offered, each product would have to work. Then, one would have to purchase and use many such products all the time. Using only one or even several would not provide meaningful protection, even if each was individually effective and even if one used that product or products all the time. Imagine, for example, that a person manages to eat nothing but the purest of foods. That deals only with hazardous substances that one might ingest when eating. It does nothing about substances threatening to enter the body via drinking and breathing. And so on.

One would have to try to shield one's body at *all* the points of access to it: Install the very finest filter below the kitchen sink so one can cook every strand of pasta in filtered water. Install a filter on every showerhead. Bathe in bottled water, like Michael Jackson. Buy furniture, carpets, drapes, and bedding made of nothing but natural materials. Use only green, nontoxic cleaning products, soaps, and shampoo. It is a toxic world, and half measures will not do.

And even then it would not be enough. As an exercise, let's go through our three points of entry to the body and consider what it would take to completely control or come close to completely controlling entry at each and every portal. This discussion will show that even with the finest products, even if one can buy them all and use them relentlessly, one can never achieve complete protection or anything approaching that.

Drinking

Bottled water and filtered water would have to be much better than they are today, offering again and again a product free of all contaminants of any significance. Then, a person would have to use bottled water and/or filtered water religiously.

We know that neither is generally true today. The quality is uneven, most likely good and sometimes better than tap water, but occasionally worse. Most consumers who say they use bottled water do so only part of the time. Only about one in five Americans say they use bottled water as their main source of drinking water. A still smaller number—one study puts it at 7 percent—use bottled water for both drinking *and* cooking.[56] If we add people who filter their drinking and cooking water, that number is considerably larger.[57]

Most people who buy bottled water, even those who buy it more than just occasionally, still also drink from the tap,[58] and even people in that 7 percent probably do not use bottled water *every* time they boil a potato. But let's not quibble. Let's consider only the extreme case of a person who drinks and cooks only with bottled water or filtered water and continue to assume, for now, that all that water is absolutely pure, free of all contaminants. (If such a person has a patron saint, it would have to be the paranoid General Jack D. Ripper in the film noir classic *Dr. Strangelove*. In one of that film's most memorable scenes, Ripper tells Mandrake, his aide-de-camp, that he drinks "only distilled water, or rain water, and only pure grain alcohol" in order to protect the "purity" of his "precious bodily fluids.") If a person drinks nothing but bottled or filtered water, is he or she protected?

Even if those bottles are filled with pure, pristine water and the filter disgorges nothing but pure water, that person is not completely protected. A. G. Ershow and K. P. Cantor found that drinking water and cooking water account for only about 60 percent of a person's total water intake.[59] The remaining 40 percent is "intrinsic" water, in part the "commercial water" added during processing as a company prepares a food item for market and in part the "biological water" naturally in food.[60] These two components of daily water intake are, it should be clear, waters that are not directly under a person's immediate control. "Commercial" water is water from the public water supply; it is drinking water, tap water. The "biological" water in foods is the water that makes crops grow; it is rainwater or river water or groundwater used to irrigate crops. It is also the water offered to farm animals.

We know, from the discussions in chapters 3 and 4, that there are hazardous substances in the water used to water crops and water livestock. A plant's roots filter out some of these substances; others are absorbed and remain in the water in the plant's tissue. Contaminants in animals' drinking water can be excreted; they can, conversely, bioaccumulate. With the one exception of the now familiar advisories to avoid eating contaminated fish and shellfish, this important pathway for waterborne contaminants moving from environment to body is largely invisible to the ordinary consumer.

So, even if bottles had in them and filters delivered nothing but the purest water, that would take care of only 60 percent of one's daily intake. Forty percent is far from trivial. If one wanted to completely eliminate any possibility that contaminants could enter the body via water ingestion, one would have to go beyond just controlling drinking water and cooking water, that is, the waters directly and immediately under one's control; one would have to try to control, somehow, the quality of intrinsic waters of all kinds, that is, again, of the water already in foods and beverages at the time they are purchased.

Eating

The organic food we have is good but not yet good enough. Organic certification not only requires that a crop be raised without application of pesticides and that the organic farmer not use certain types of fertilizers, but also that the soil has not been sprayed with pesticides for three years, and that the organic farmer makes efforts to protect his crop from pesticide drift and other nonorganic materials that might migrate from adjacent fields. However, we recall that DDT, banned for decades, is still found in soil.[61] And organic certification does not specify that the water applied to the crop be treated until it is pure, or even that it has to be water in compliance with the standards of the Safe Drinking Water Act. If we wished to ingest no contaminants when eating—and that includes contaminants in the water intrinsic in food, 40 percent of daily water intake—we would have to insist on what might be thought of as a *hyperorganic* diet. One would require crops grown in a completely organic environment, crops that are not only not sprayed with pesticides and not grown with applications of chemical fertilizers, but crops grown in soil that is completely free of contaminants, and watered only with purified water. Also, if at all possible, one would require crops that are shielded from the wind. One would require cows, pigs, and chickens to be fed not only hyperorganic crops but also to be given only pure water to drink. Soft drinks, juices, and beer would have to not only be made with nothing but organic ingredients, but the "commercial" water added to them or used in making them would have to be pure too.

Breathing

An individual could, in principle, eliminate one's exposure to contaminants in indoor air. It would require a set of clear state and federal standards for nontoxic building materials, home furnishings, household products, and personal hygiene products. Then one could with some assurance build a nontoxic home, fill it with nontoxic furnishings, use only natural products to clean one's home, and apply only organic or natural or nontoxic products to one's body—if one had the money.

Outdoor Air: An Absolute Limit on the Inverted Quarantine Approach

The discussion so far leads me to conclude that at present the inverted quarantine approach to preventing environmental contaminants entering our bodies does not work very well, principally because the products individually offer limited and uncertain benefits, and secondarily because I think few people can afford to buy the whole array of goods that would be needed, even if each product did what it was supposed to do. For the strategy to work better, two things would have to happen: the products would all have to get better, and they would have to be more affordable (priced similar to the products' conventional counterparts). We are not likely to see either happen any time soon.

Assuming for the moment that all those things did happen, inverted quarantine, even at its theoretical best, runs up against something like an absolute limit when it tries to deal with contaminants in outdoor air. Outdoor air (which, since there is constant exchange between outdoor air and the air in one's home, is breathed not just outdoors) poses what I believe are insurmountable challenges to the inverted quarantine approach.

Home location does help. It is better not to live next to or directly downwind from toxics-emitting industrial facilities. Diesel exhaust, recent research shows, is a serious health hazard,[62] so it is better to live some distance from major transportation corridors and warehouse districts. No wonder environmental justice research has shown that those who can afford to buy homes far from such urban and industrial landscapes do so.

Outdoor air, though, has suspended in it more than just pollutants from local sources. Materials can travel long distances in air, crossing oceans and continents, halfway around the globe and even farther.

It is been known for several hundred years, of course, that modern industrial production generates immense amounts of air pollution. Recall the poet William Blake's description of factories as "dark Satanic mills," bellowing blackened air from their smokestacks, casting a pall over the city, sickening the inhabitants. But the impacts appeared then to be largely localized. It took years, and the slow accumulation of evidence, for us to understand just how far pollutants could travel.

Radioactive fallout from atomic testing showed very dramatically how far winds can carry pollutants. Invisible poison could fall from the sky hundreds of miles from the actual point of detonation. Even if you did not see it, it could make you very ill. For years it had been rumored that fallout from aboveground atomic tests in Nevada had resulted in an increased incidence of cancer in downwind communities. In 1997, the National Cancer Institute confirmed it was true. People living downwind of the test site in Nevada did turn out to have elevated rates of thyroid cancer. Five years later, a larger follow-up study by the Centers for Disease Control and Prevention showed that people living in states immediately east and northeast of Nevada got the highest doses, but also found that almost every American, no matter where they were living in the continental United States, was exposed to some radiation from the tests.[63]

Because so much of the relevant data were classified, it took several decades to definitively prove that fallout from atomic tests could harm people's health many miles away. Meanwhile, other scientists were slowly amassing evidence that other sorts of environmental chemicals were traveling long distances, endangering people's health far downwind.

In the 1960s, Rachel Carson assembled the evidence that showed that pesticides traveled far, by air and by water, downwind, downstream, wreaking destruction far from where they were first applied to crops.

In the 1980s, scientists brought to the public's attention the emerging problem of "acid rain." The water in many lakes in northeastern United States was becoming more acidic, eventually to the point that the lakes' ecosystems threatened to crash. For us, the important point is the discovery that the rain in New England had become acidic because of power plant emissions in the Midwest.

In a separate development, also in the 1980s, Canadian researchers were measuring the level of PCBs in the breast milk of nursing mothers in Montreal. They thought they should have a control group to whom the test subjects in Montreal could be compared. Seeking a pristine population, they headed north. Some Inuit women, seemingly ideal controls because they lived thousands of miles distant from where PCBs were manufactured and used, agreed to participate in the study. The researchers were surprised to find high levels of PCBs in the breast milk of the Inuit mothers, far higher than in the breast milk of the Montreal mothers. Eventually, the following answer was worked out: Global air circulation had taken persistent organic pollutants, such as PCBs, toward the poles. When the air got cold enough, the chemicals precipitated out and returned to earth, to the sea. There they were taken up by smaller organisms, which were, in turn, eaten by fish that were themselves eaten by seals and whales, animals at the top of the arctic food chain. As these persistent organic pollutants worked their way up the food chain, the processes of bioaccumulation and biomagnification produced very high levels of PCBs and other persistent organic chemicals in the fatty tissue of seals and whales. Because the hunting peoples of the north eat seal and whale, the persistent organic chemicals, produced and used thousands of miles to the south, ended up in the bodies of the Inuit. Later work showed Inuits' bodies had not only PCBs but also a host of other persistent organic chemicals, such as chlordane, toxaphene, and dioxins, and toxic heavy metals such as lead, cadmium, and mercury.[64]

These days we are learning that Asia's economic miracle is bringing us more than just cheap goods. Environmental regulation in China lags far behind Europe and the United States. Weaker environmental regulation is one of the reasons American manufacturing has been moving to China (low labor costs is, obviously, another). It turns out, though, that shifting factory production to China does not fully export the environmental burdens of industrial activity to that nation. It does do that to a great degree—eyewitness accounts describe appalling levels of air pollution in China's industrial cities. At least some of the waste products spewed from Chinese factories, though, rise high enough to catch the wind, cross the Pacific, and end up in the United States. Just taking note of the occasional news story, certainly not undertaking anything like a systematic search, I have seen reports of ozone, fine particulates, mercury, arsenic, copper, silver, and zinc generated by industrial activity, such as coal burning and metal smelting, in China, which then cross the Pacific Ocean, pollute the ocean, and pollute the air over Hawaii, California, and other places in the United States.[65]

Every few months a new piece of research shows, again, how materials produced, used, and discarded in one part of the world find their way to distant lands. The distance traversed can be regional, from Ontario to Rhode Island, Germany to Sweden.[66] It can be global: dust from the Sahara blows almost all the way around the world, until it is deposited in the ocean off West Florida. There, the iron in the dust fuels toxic algae blooms.[67] Pollutants move both latitudinally, from the equator toward the poles, and longitudinally, from west to east. As Thomas Cahill, physics professor emeritus at the University of California, Davis, put it, "We live in a small world. We breathe each other's air."[68] The director of the nonprofit Clean Air Task Force says, even more succinctly, "the globe is one air shed."[69]

Living in a nice residential neighborhood, away from the chemical plant and the industrial park, away from all that truck traffic, will definitely protect one from some, perhaps a great deal, of pollution from local sources. But think of the Inuit mothers. Who would have imagined that they would have PCBs in their milk glands? The expensive, leafy suburb may seem far removed from local industrial activity, far enough away to be insulated from its effluents, but it does not seem so far away compared to Inuit villages north of the Arctic Circle.

We have reached something like an absolute limit here. We cannot not breathe. There just does not seem to be a practical way to make a personal membrane impermeable to pollutants in ambient, outdoor air. One would have to wear something like the spacesuits used by NASA astronauts, with a self-contained air supply; or something like a scuba diver's breathing apparatus; or a portable version of the personal biosphere built, once, in the 1970s, for that little boy who was born without an immune system and who stayed alive only because his parents built for him an immense bubble, a pristine home within their home.

Probabilities, Costs, and Limits

Does environmental inverted quarantine work? There is no one answer to that question. There are a range of answers, depending on what products a person uses and how many. Use one or two natural items of uncertain quality, and you get little or no real protection. Eat organic foods most of the time and avoid living downwind of toxics-emitting factories, and you get some, possibly substantial, protection. Go further: buy a house in the hills, outfit it with an array of filters, eat only designer organic food, shun furniture made of particleboard and carpet made of synthetic fiber.

You will get even better, though far from complete, protection. The more time, effort, and money devoted to the effort, the higher the level of protection.

As the number of items grows, the cost soars, so the income gradient is necessarily steep. A middle-class family can afford not to live near major highways or next to oil refineries and chemical plants. That family can probably also afford to eat organic, at least some of the time. It would take real money, much more disposable income than most middle-class families have, to implement the whole program.

At its most intense, inverted quarantine exacts a psychic toll, as well (similar to the existential costs I described at the end of chapter 2). Practicing inverted quarantine on such a scale requires that one organize one's whole existence around the perceived need to erect and forever maintain barriers through constant and repetitive acts of monitoring, avoidance, separation, and enclosure. Inverted quarantine would have to become the organizing principle for the conduct of everyday life. It would be the life of a person who suffers from MCS, Multiple Chemical Sensitivity syndrome, and organizes her or his whole existence around the task of avoiding contact with anything but natural materials. Those who suffer from MCS have no choice. They must live this way. Is it possible that a person can choose to live this way voluntarily? Such perpetual vigilance borders on the clinical, on phobia and obsession. This is a life indistinguishable from the life of someone who suffers from obsessive-compulsive disorder. Inverted quarantine becomes voluntary self-imprisonment.

And no matter how much one spends or how obsessive one is, two facts will keep one from ever achieving anything approaching complete protection. First, all the items bought and used would have to work. As we have seen, the actual items on the market are very far from that. Second, one has to breathe. I do not believe there is a practical solution to the outdoor air—"the globe is one airshed"—problem, unless one is willing to make the ultimate sacrifice, completely give up freely moving about in the world, retreat into an actual bubble space outfitted with powerful air scrubbers or an artificial air supply.

7. Political Anesthesia

Inverted quarantine products do not work all that well. That does not concern me much, though it might not be welcome news to those who believe in these products and spend good money for them. What matters to me more is that people *think* they work. As the great American sociologist W. I. Thomas wrote, long ago, "If men define situations as real, they are real in their consequences."[1] In this chapter, I try to understand what, in this instance, those consequences might be.

I first started thinking about inverted quarantine when some years ago I walked into a supermarket and was amazed to see shelf after shelf filled with bottled waters. Bottled water became for me not only the first but also, provisionally, the paradigmatic case of the inverted quarantine response to environmental threat. It was the first example I studied in depth. I learned, among other things, that the quality of the product was suspect, as I reported in chapter 6. I also learned that just about everything about bottled water production and consumption is bad for the environment. It does not do away with toxics; it just shifts toxics from one location to another, now more concentrated than before. It consumes resources and energy, then leaves behind mountains of plastic bottles.

Generalizing (prematurely, as it turns out), I thought I had it: inverted quarantine products do not work, *and* when large enough numbers of people use them, it ends up, paradoxically, trashing the environment further.

I thought I had the answer, and an attractively ironic one, to boot. Unfortunately, my second case study, of organic foods, convinced me that it is not always and necessarily so, that mass use of inverted quarantine products does not have to end up degrading the environment in that direct sense. I concluded, reluctantly, that if I were to find unintended consequences, I would have to look elsewhere.

After more research and reflection, I concluded that inverted quarantine does indeed harm the environment, but it does so in most instances in a more indirect, more circuitous manner. Doing inverted quarantine changes people's *experience*. It alters their perception of their situation. Their sense of being at risk diminishes. The feeling, correct or not, that they have done something effective to protect themselves reduces the urgency to do something more about what, until then, felt threatening to them. If many people experience such a reduction in urgency, that will have consequences in a democracy, in a society where mass sentiment affects what government does. I think this reduction in urgency, this sense of having taken care of the threat, produces what we might call "political anesthesia."

I will make the argument that political anesthesia is *the* important unintended consequence of mass practice of inverted quarantine presently. Rather than go directly to that, though, I would like to first recount how I arrived at my initial, wrong answer and why I later had to reject it.

The Wrong Conclusion: Inverted Quarantine's Physical Impacts Are the Problem

Bottled Water's Physical Impacts

Consider the whole arc of the product cycle, from extraction, to processing, to bottling and distribution, to postconsumer waste. At every point, the environment is degraded.

Extraction

If the water in the bottle is drawn from a spring, a remote and pristine bit of nature has to be developed, *industrialized*. Millions of gallons of water have to be captured. That requires pipes, pumps, and holding tanks. Essentially, a small factory has to be built and operated at the source, either to bottle the water right there or to collect it and store it until it can be trucked to another location for bottling. Roads have to be widened, or new roads built where there were no roads before. A steady stream of trucks carries water from the site, making noise, belching diesel exhaust.

Purification

If the water in the bottle comes instead from a utility, it has to be treated some more before it is ready for its new identity. Regardless what treatment is used, filtration, reverse osmosis, ozonation, or distillation, that treatment requires equipment and energy. Two streams flow from this process: a stream of the desired commodity, treated water; and a stream of waste products, including substances that have been removed from the water, used filters, and processing equipment past its useful life. The first stream heads to market; the second has to be disposed of in some fashion.

Packaging

Billions of plastic bottles have to be made to hold all that water.[2] Total demand for oil increases. Petrochemical refineries and plastics companies turn that oil into a variety of plastic resins. The resins are blow-molded to make bottles, which are then shipped to bottling plants. Making bottles for water requires no new or special industrial processes. Even if there were no bottled waters on the market, billions of plastic bottles would be made each year, for soft drinks, milk, ketchup, peanut butter, and many other food items, and also for household cleaning products, motor oil, and so on. Putting water in bottles just increases the volume, the sheer number of bottles that are made each year. That increment requires the manufacture of *more* plastics, *more* truck traffic from container maker to bottler, more energy at every step.[3] We are talking about large quantities of product, billions of bottles, by one estimate 1.5 million tons—3 billion pounds—of plastic to hold all that water.[4]

Transporting to Market

Once bottle and water are united and the product is labeled and boxed, it joins, and swells, the many other streams of food and beverage items traveling from manufacturer to supermarkets, convenience stores, and restaurants. Transporting all those bottles to market means more packaging, more product in warehouses, more forklifts loading pallets stacked with cardboard boxes into trucks, more truck traffic.

Postconsumer Waste

And in the end, there is the empty bottle. To its credit, the industry strongly encourages recycling. Something less than 25 percent of PET and HDPE bottles are recycled,[5] very respectable when compared with the recycling rate for most other food and beverage packaging. The relevant comparison, though, is not to other packaging but to water from the tap, which creates no comparable postconsumer waste issue. More than three empty bottles out of four are thrown away. They end up in landfills. A 2002 study funded by Coca-Cola, Waste Management, Inc., and others estimated that 114 billion beverage containers are thrown away annually.[6] Picture a mountain made of a hundred billion plastic bottles. Those are the bottles Americans threw away last year. Then picture another mountain next to it. That is the mountain we are building this year. Consumption is still growing, so you have to imagine that this mountain is somewhat bigger than the first. Picture, then, a whole mountain range, ten, twenty, fifty peaks, a new peak each year, each taller than the one before.

The Decisive Counterexample: Beneficial Impacts of Organic Agriculture

Just about everything about bottled water production and consumption is bad for the environment. I thought I had my argument nailed. Then I turned my attention to organic food and learned that I had gotten it wrong.

To make a bottle full of (more or less) clean water, mountain springs are turned into industrial sites, plastic polymers have to be produced, energy used, hazardous wastes and postconsumer, "solid" wastes generated. The production process transforms nature, pollutes nature. Clean consumption is compatible with, even requires, dirty production. It is an amazing contradiction.

In contrast, in the case of organic food, dirty production cannot make a clean product. A farm has to practice organic agriculture if it is to market its crops as organic. If the farm wants to sell organic produce, it has to *grow* that crop organically. If a ranch wants to market its meats as organic, the animals cannot be injected with hormones or fed antibiotics. Unlike with bottled water, in the case of organic foods *the conditions of production cannot be separated from the qualities of the final product.*

But we should not idealize organic agriculture. Organic is still tiny, only 2 to 3 percent of the total food market, but because it is growing so fast right now, big food corporations have been buying into the organic segment of the market. These new actors, likely the future face of organic, do not embrace all the values and practices of the more "traditional," more explicitly political organic farmer, values such as a commitment to small, family farms, growing crops locally, and treating agricultural laborers well. In many ways, their methods resemble those of conventional agriculture.[7] They have been accused of trying to manipulate or "water down" recently instituted FDA organic standards.[8]

Granted all that, it seems to me that even in its organic-evolving-toward-industrial-agriculture manifestation, organic agriculture is beneficial for the environment. The more acreage farmed without the use of chemical pesticides and fertilizers, the better for the local environment. As I discussed in some detail in chapter 3, the pesticides and nutrients applied to conventionally grown crops are a major source of local surface water and local groundwater contamination. Furthermore, some of those chemicals sprayed on crops leave the place where they were applied and migrate via air, surface water, and groundwater to other places. So growing crops organically is good not only for the immediate area; it means a lower toxic load for the environment as a whole.

Organic agriculture forced me to conclude that mass flight to environmental inverted quarantine products does not always and necessarily harm the environment. If environmental inverted quarantine had characteristic, inherent, necessary adverse long-term impacts, I had not yet found them.

Political Anesthesia: Individual Acts of Inverted Quarantine

A Clue from Bottled Water

I reexamined my materials and found a clue in what had initially seemed to me a rather marginal set of notes buried in the stack of my bottled water materials. In 1999, the Natural Resources Defense Council (NRDC) published a major study of bottled waters. The NRDC found that the quality of bottled waters varies considerably, that many samples had some contaminants in them, and some samples even failed to meet basic Safe Drinking Water

Act standards. The NRDC argued that "the long-term solution to our water woes" was not to rely on bottled water but "to fix our tap water so it is safe for everyone."[9] This was the response from www.bottledwaterweb.com, the Web's self-declared "Definitive Bottled Water Site": "Fix our tap water so it is safe for everyone? This will never happen." Why? Because "municipalities [are] using antiquated technologies" and the "infrastructure of the municipal systems [is] deteriorating."[10] The cost of fixing all that would be "astronomical." It is "a pipe dream," bottledwaterweb.com said, that society would ever be willing to spend that kind of money.

Coming from the bottled water industry, such prophecy may not just be self-serving, I thought; it could also be self-fulfilling. Maybe bottledwaterweb.com was right. Maybe the money would never be spent. But what, exactly, was the cause-and-effect relationship here between money not spent on infrastructure and consumption of bottled water? According to the industry, the sequence was as follows: tap water is suspect; money will never be spent to improve it; therefore people will continue to drink bottled water. What if the actual sequence went instead like this: tap water is suspect; people switch to bottled water and stop worrying about tap water; political support for spending on infrastructure weakens; money is not spent.

I looked at some documents and found that bottledwaterweb .com was right—the costs are indeed astronomical. The physical infrastructure of the public water system—the purification plants, the miles of pipes that carry the water from treatment plants to consumers—is aging. Many billions of dollars will have to be spent over the next twenty years just to keep the current system operating as well as it is operating today, just to repair or replace parts of the infrastructure as they wear out or fail.

Those first billions will only pay to keep the current system in good repair. That is only the beginning. If some of the substances not currently regulated are eventually shown to pose a significant threat to public health, new treatment technologies will have to be installed to remove those substances from the water before it is delivered to consumers. Sheer population growth will require, in any case, that every part of the infrastructure be expanded to keep pace with increasing demand.

Estimates of what it would cost to keep the system in good repair, keep up with increasing demand, and upgrade treatment to deal with new contaminants start at 150 billion dollars over the next twenty years. That is the low estimate. In 2002, the EPA projected that there would be a $534 billion gap over the next twenty years between the amounts that would be needed to fully fund the activities mandated by the Clean Water Act and the Safe Drinking Water Act and the amounts then budgeted for implementing those acts. The Water Infrastructure Network puts the amount at twice that, $1,000 billion—a *trillion* dollars.[11]

The federal government is currently spending only a tiny fraction of that, perhaps 2 billion dollars per year, on sewer and drinking water projects.[12] If funding holds steady, that will add up to $40 billion over the next twenty years, far short of even the low-end estimate of future needs.

Spending is unlikely to grow anytime soon. The Bush administration is deeply hostile to environmental protection. It has tried to cut funding for the Clean Water Act and the Safe Drinking Water Act, not to increase them.[13] I have not heard a single nationally prominent politician say anything about spending another $500 billion over the next twenty years on water projects.[14] As I write this, in fall 2006, tax cuts and war spending have pushed federal deficits to, or near, record levels. Billions more will have to be spent to rebuild our Gulf Coast communities, devastated by the hurricanes of 2005. If and when those huge but time-limited items are dealt with, there is the long-term problem of the tens of millions of "baby boomers" who are approaching their golden years. Just paying for their health care will put tremendous strain on government budgets. There is, on the other hand, little enthusiasm for tax increases.

In such circumstances, politicians would be willing to consider spending hundreds of billions of dollars for water quality only if they believed that public support for it was both broad and deep. Polling data do show that support *is* broad. By large majorities, Americans say they are concerned about water pollution and that they support strengthening regulations to ensure that rivers and lakes are cleaner and the drinking water supply safer.[15] We need to remember, though, the important distinction opinion pollsters make between issue support and issue *salience*. Although typically around 80 percent of Americans say they support stronger standards, only 22 percent of voters say that candidates' positions on environmental protection are a factor in who they vote for. "Focus groups ranked the environment last out of nine issues tested, in terms of both its impact on their vote and personal importance. The environment fell behind the economy and jobs, health care, Iraq, Social Security, terrorism, education, moral values, and taxes."[16]

How does one explain this gap between caring about water quality in the abstract and caring about it enough to make a difference in how a person votes? I think that mass belief in inverted quarantine is part of the answer. Tens of millions of citizens believe they have successfully insulated themselves from the problems of the public water supply because they drink bottled water or because they filter the water coming into their homes. They continue to care about water quality, but is there any real motivation left to do anything more about it?

Admittedly, using bottled water (or filtering tap water) and politically supporting government spending to clean up *everybody's* water are not mutually exclusive behaviors. There are, I am sure, many Americans who consider themselves liberals and are politically active who also eat organic food, drink bottled water, and so forth. But I would suggest that in the vast majority of cases, when one believes that one has successfully shielded oneself from a threatening condition, that threat's salience—in this case the salience of tap water quality—is diminished, and in many cases disappears completely. The act of inverted quarantine produces a subjective sense of security. You have dealt with that problem. You do not have to think about *that* any more. Naturally enough, you think and worry about other things, social problems that you do not think you can do anything about as an individual, personal problems, all the many things one needs to attend to every day, problems concerning work, family, money.

Belief in the efficacy of the inverted quarantine solution to one's toxic fears, though largely unwarranted, translates into the subjective state I think deserves to be described as a form of anesthesia. What is the likely consequence when a significant fraction of the public imagines that they have successfully bought their way out of a collective problem? The political system is sensitive to what upsets people versus what does not. Why would a congressman or a senator vote to spend hundreds of billions of dollars for massive public works if voters in their district—the most affluent and politically most savvy voters in their district—are *not* calling them or writing them because those voters believe they have solved that problem on their own, individually. No significant political demand from citizens means no pressure on elected officials. The money needed to keep the infrastructure in good repair is not spent. Newer contaminants are not regulated. Drinking water quality deteriorates over time.

The New Hypothesis

Here, then, was a second, more indirect way bottled water use has consequences for the environment. Yes, billions of bottles of H_2O directly, physically degrade the environment with higher energy use, dirty used filters that have to be landfilled, mountains of empty plastic bottles, and so on. In addition to that, though, because so many consumers believe that drinking bottled water in lieu of drinking tap water will protect them from toxics, the political support that is needed to address water quality issues more substantively, systemically, just is not there.

The first kind of impact, direct physical damage to the environment, turns out not to generalize well to other cases. The second kind of impact, a false sense of security undercutting political support for reform, I believe does generalize and is, in fact, the unintended consequence that really matters.

Does Consumption of Organic Food Foster Political Anesthesia, Too?

Bottled water harms the environment. Belief in bottled water and water filters drains the support that is needed to get government to spend hundreds of billions to maintain and improve tap water quality. Faith in bottled water in effect helps perpetuate the conditions and processes that have made the public water supply suspect. We have already seen that organic food is innocent of the first charge: it is beneficial for the environment, not harmful. What about the second charge? Are the political consequences of organic food consumption radically different from the consequences of mass consumption of bottled water and water filters? Is organic food again the decisive counterexample that will force me, once again, to give up a generalization about inverted quarantine's unintended consequences?

It might seem so. Organic food's most ardent supporters say that buying organic food is for them a political act. When they buy organic food, they are not just trying to shield their bodies from harm. They are helping create a viable alternative to conventional agriculture, an alternative that does not require using toxic chemicals, that preserves and enhances soil quality, that has a more benign impact on biodiversity than conventional agriculture does. They believe that consuming organic food and supporting organic agriculture are ways to participate in the larger movement to preserve the environment. Organic food consumers are changing the world one shopping basket at a time.

Other consumers of organic food do not see themselves as environmental activists exactly, but they are generally sympathetic to green and progressive causes. As they shop in the organic food store and read the labels and the posted information, such folks spend time in the "discursive space" of the environmental movement, exposed to its worldview and ideology. Product labels describe the conditions under which the ingredients in the product were grown. Labels inform the customer that a certain percentage of the profits are being donated to various environmental causes. Enter one of the natural food stores in my town, and you travel through time almost back to the Summer of Love, or at least back to the heyday of the alternative, co-op movement of the early 1970s, complete with the music, the posters, and a dreadlocked "hippie" wearing a tie-dyed T-shirt working the checkout counter. You have to be somewhat identified with the counterculture just to feel comfortable walking into the place. The other organic food market I shop at looks physically much more like your average, if smallish, supermarket. What it lacks in atmospherics, it makes up for by being even more overtly political than the more hippie-ish store down the street. The fish and seafood selections are not only marked for price but are also color coded to show how harmful the harvesting of that specific species of fish is to the health of the ocean. Signs also tell consumers if a fruit or vegetable was grown locally. Signs next to the checkout stand tell the consumer that part of the store's profits go to community groups and local social movement organizations. On the way out the door, there is a rack filled with movement publications.

If these depictions of organic food consumption are accurate, the political impact is completely the opposite of the "buy bottled water; feel secure; stop caring about everyone else's water" effect described earlier. Consuming organic is either an expression of pre-existing left political leanings or is itself a politicizing experience. Because production and consumption of organic food are so explicitly political, they appear to be the toughest possible test of the inverted quarantine–political anesthesia thesis.

I believe, though, that in this case organic food consumption will not serve as a valid counterexample. It will not undermine the anesthesia thesis. To see why, we have to recall, from chapter 4, that as organic food consumption has changed from being a marginal, somewhat cultish practice to becoming an acceptable, even mainstream choice, the nature of the typical organic food consumer has changed. Remember the distinction Michael Pollan draws between two types of organic consumers, "true naturals" and "health seekers." The "true natural" is "a committed, activist consumer. . . . socially conscious . . . devoted to the proposition of 'better food for a better planet.'" The "health seeker," typically "more affluent," is "more interested in their own health than that of the planet. They buy supplements, work out, drink wine, drive imported cars. . . . The chief reason the health seeker will buy organic is for the perceived health benefit."[17] Health seekers, in other words, are the inverted quarantine consumers of organic food. They are not particularly interested in transforming all of agriculture, because all that matters to them is that they have taken measures to protect *their* bodies from harm.

Health seekers already outnumber true naturals by more than two to one.[18] Market research conducted by some big agribusiness corporations that are moving into the organic sector forecasts that that gap will continue to grow. There are not that many more "true naturals" out there, the market research suggests, compared to a much larger pool of potential "health seekers" who have yet to go organic.

The health seeker is by and large indifferent to the plight of ecosystems assaulted by the impacts of conventional agriculture and indifferent to the plight of others who cannot afford to make the same choices they do and who must continue to eat food that is conventionally grown. Believing that they have successfully taken care to shield themselves from harm, their consumption choices lead them not to activism but its opposite, political inattention and indifference. In the end, the anesthetic effect is no different than it is for bottled water and water filters. Once again large-scale flight to inverted quarantine has the effect of removing from the political stage a substantial number of savvy, influential people who might otherwise support reforming how *everyone's* food is grown.

But wait. Motivation, or intention, or the "consciousness" of the consumer may not matter. From the point of view of the environment, what does it matter if the consumer is a true natural or a health seeker? If consumption of organic foods goes up, the number of acres farmed organically grows. Agriculture is gradually, incrementally transformed over time. The environment benefits.

The flaw in that line of argument (which would otherwise be worth considering) is that growth in consumption of organic food, currently still impressive, is likely to eventually slow considerably, perhaps completely stop—stop, in fact, well short of the point where it would begin to really transform how most food is grown. Two facts interacting with each other will, I believe, eventually retard the growth of the organic market: the high price of organic foods and increasing income and wealth inequality in society.

Prices for some organic foods have come down and are roughly comparable or only slightly more expensive than their conventionally grown counterparts. Other organic items, including important components of a typical diet such as meat and dairy, are more expensive and sometimes a great deal more expensive than conventionally grown counterparts (see chapter 4). Only a small minority of Americans have the kind of income that allows them to choose an all-organic diet. Even for someone with a decent middle-class income, it can be too expensive to consistently maintain an all-organic diet. One has to pick and choose—maybe pay extra for organic milk so the kids don't drink milk from cows injected with bovine growth hormone, or buy organic fruits and vegetables so the kids' bodies will not bioaccumulate pesticide residues quite so fast. But it can get expensive fast if one decides one must have every food item be organic every day. And there is a line, a level of income, below which a family cannot ever make the organic choice. Other, necessary things must come first.

Consider next the likely impact of current trends in wealth and income inequality in the United States. Income and wealth inequalities decreased during the first half of the century, were generally flat during the 1960s and 1970s, then began to grow steadily after about 1980 or so. Between 1952 and 1982, the top 10 percent's share of national *income* (note: these figures exclude income from capital gains) held steady in the 32 to 33 percent range; by 2000, the top 10 percent's share of national income had climbed to 42 percent. From 1962 to 1982, the top 1 percent's share of national income

. (again, excluding income from capital gains) hovered around 8 percent; by 2000 it had risen to almost 15 percent.[19]

More recent income data published by the IRS show these trends continuing. Between 2002 and 2003, individuals in the top one-tenth of 1 percent increased their income—in one year—by 9.5 percent. The rest of the top 1 percent increased their income by 3.7 percent. All other individuals, the other 99 percent, saw their income go up about 2 percent. Since the rate of inflation that year was 2.3 percent, real income fell for everyone except those in the top 1 percent.[20]

That is income. Turning to *wealth,* the share of total national wealth held by the top 1 percent of households was at about 20 percent in the late 1970s. In the next twenty years it would almost double. By 1995, the top 1 percent had more than 38 percent of the nation's wealth.[21]

Consumption of organic foods will continue to grow, but that growth cannot continue much longer unless the price of organic foods falls to where it does not cost any more (or only negligibly more) than conventionally grown food does, or trends in wealth and income inequality reverse and those inequalities begin to diminish, or both. Neither is likely to happen.

Economies of scale and new methods of organic farming (that is, methods more like conventional agriculture, minus the chemicals) may drive down the price of some organic food items somewhat. One has to remember, though, that conventional agricultural production methods, spraying crops with chemical pesticides, feeding farm animals hormones and antibiotics, raising those animals in what amount to animal concentration camps, were adopted exactly because they lowered the price of production, making it possible to market foods at low prices and still make a decent profit. For many items, the cost of producing them organically is intrinsically higher and is not likely to fall to the point that that item comes to be priced competitively with its conventionally grown counterpart.

As long as organic foods cost more, the class dimension of consumption I have described will not change. Only a tiny minority will ever be able to afford an *all*-organic diet. In the middle, a bigger group will be able to afford to have some organic items in their diet. Many will continue to be able to afford only conventionally grown foods. If prices do not equalize, demand will not keep going up unless the twenty-five-year trend of increasing inequalities in wealth and income is reversed. I do not see any social or political process on the horizon that can drive any such reversal. Income and wealth inequalities will either remain unchanged or, more likely, will continue to grow larger for the foreseeable future.

That means that the number of people who are for financial reasons more or less completely shut out of the organic food market is not about to fall. Those numbers may, in fact, rise as incomes stagnate, as the cost of health care continues to rise far faster than the overall rate of inflation, as the cost of housing continues to rise, as the price of a tank of gas, the cost of heating one's home in winter, and the price of other necessities continue to rise.

That is why I believe that growth in organic food consumption will slow and stop long before it reaches the point where most of agricultural production or even a significant fraction of it is transformed. The toxic mode of agricultural production will not simply wither away. The most likely outcome, I believe, is not that organic food consumption will gradually transform agriculture, but that we will have two agricultural systems side by side: a large conventional sector that grows affordable, if slightly contaminated, foodstuffs for the majority, and a smaller one producing organic alternatives for a minority, largely made up of affluent health seekers.

When demand for organic food stops growing, market forces will cease to be the engine for further transformation. At that point, if further progress were to be made, it would have to be political activism that pushes things along. This is where the motivation, or "consciousness," of the organic consumer comes to matter again. It will matter that health seekers outnumber true naturals by more than two to one and that their numbers continue to grow faster than true naturals'.

The environmental impacts in the end will be better than the impacts of bottled water consumption. Because there is a market for organic foodstuffs, some farm acreage has no chemicals applied to it, and some farm animals are raised without hormones and antibiotics. But because health seekers already dominate the market for organic and their dominance will only increase, it seems to me that no matter how paradoxical it may at first sound, at some point consumption of organic foods will turn from being a force for transforming agriculture into the opposite of that, into a force that impedes further progress and helps actually perpetuate the continued dominance of conventional, chemicalized agriculture.

Remember, health seekers are "more interested in their own health than that of the planet." Again, these are the more affluent, more politically savvy members of society. Once they feel safe, they are not likely to spend a lot of time and energy mobilizing for general agricultural reform.

For these reasons, it seems to me that here again, in the most explicitly politicized example of the phenomenon we have been examining, the mass practice of inverted quarantine will not end up sustaining a movement for change but will instead produce the opposite of that, political anesthesia.

Other Inverted Quarantine Products

I believe this analysis of the eventual consequences of organic food consumption also works for other products I have described.

Consider, say, nontoxic household cleaning products, or personal hygiene products that do not contain phthalates, or furniture made of whole wood rather than particleboard that off-gasses formaldehyde. To the degree that they capture a fraction of their respective markets, the environment will carry less of a toxic load. Although I have not been able to find much market research (some on growth trends; nothing about consumer motivation), it is likely that here, too, some consumers are politicized true naturals, while others are only health seekers. And such products are typically more expensive than their conventionally made counterparts, so income inequalities limit market penetration. In all these ways, these products are very much like organic food.

The outcome is likely to be similar too. An affluent minority—a savvy and influential minority whose political influence is disproportionately greater than their numbers—buys out of the toxic environment, believes it has taken care of the problem for themselves, and loses further interest in that particular toxics issue. Support for more substantive reform weakens. At best, as with organic foods, the situation will tend toward the creation of a permanent dual market, the larger of which consists of products manufactured in a toxic work environment and that contain toxic ingredients, and whose production and consumption continue to discharge these substances into the environment.

Political Anesthesia vis-à-vis the Toxic Mode of Production

Whether it be toxics in the water, in food, or in other consumer products, the availability of seemingly safer alternative products reduces the pressure to deal with that threat to one's health more systemically. I wish to go a step further now and argue that the anesthesia effect goes beyond this or that individual toxic threat. It impedes the development of public sentiment that would support a broader reconsideration of the toxic mode of production in general.

Let's return to our earlier discussion of the distinction students of public opinion make between support and salience. Since we know that threats to one's health or to the health of one's children matter a great deal and we know Americans have been bombarded with information about toxic threats, we are not surprised that opinion surveys show that Americans are "concerned." Overwhelmingly, we say we want clean air and clean water. At the same time, though, the surveys also show, as I have already described, that most Americans do not seem particularly *intent* that something substantial be done about these issues. Support for environmental protection is wide but thin. Environmental issues have substantial support but low salience. We are concerned about these issues, but we rank a substantial number of issues as more important.

We are concerned, but not enough to have that sentiment determine how we vote. That curious absence of salience is demonstrated not only by opinion survey results. It also shows in the lack of any significant backlash against officeholders who advocate deregulation. In 1969 and 1970, the federal government passed a series of new, landmark regulatory laws aimed at cleaning up air and water and protecting the health of citizens, workers, and consumers. Only a few years later, conservatives began to organize a campaign to weaken and, if possible, repeal those regulatory initiatives. The antiregulatory campaign saw its first major victories in the early years of the Reagan administration.[22] Today, antiregulatory forces control federal policy. The agencies that are supposed to implement environmental laws, such as the EPA and the Department of the Interior, are run by administrators hostile to their statutory mandates. Republicans in the House and Senate support legislation that would weaken the Clean Air and Clean Water acts. If these issues had real salience, there would be a political price to pay for supporting policies that weaken regulations, thereby exposing the public to greater hazards.

How are we to understand this gap between support and salience?

I believe that the availability of inverted quarantine products is responsible for much of it. Imagine, for a moment, a world in which no inverted quarantine products were on the market (or people did not believe that they work and did not buy them). When a person learned about toxic threats to their health and their children's health, there would be no individual escape hatch, real or imagined. Resignation or "doing something" would be the only options.

Issue salience would no longer be as much of a problem. Citizens would be more adamant about the need for regulations that reduce the amounts of toxics in air, food, and water.

And from that initial, immediately felt experience of a threat to one's own health, support for more general environmental reforms could grow. At least some, perhaps many, would begin to think that the problem was not just this or that toxic product. From the seeds of personal, bodily vulnerability could grow, organically, so to speak, an understanding that more systemic changes were necessary.

This is not just some ungrounded fantasy. It is directly analogous to processes I documented in my book *EcoPopulism*.[23] In that book I described how toxic threats to one's health can grow into much more encompassing forms of political awareness. The grassroots toxics movement was made up many local organizations that started either because the members' community had been contaminated by an uncontrolled release of hazardous wastes (industrial wastes improperly disposed decades earlier) or because a company was proposing to site a new hazardous waste facility, such as a landfill or an incinerator, in their community. Scared, and also angry, residents felt they had to do something. They talked with neighbors. They started to organize.

At first, they had a very limited sense of what was wrong and what needed to be done. The problem was this one spill or this one proposed landfill. It was this one corporation, which had not taken proper care disposing its wastes. Or it was that waste company with the poor track record seeking a permit to build a new waste treatment facility. The solution was to win *this* specific fight. In other words, when they first got involved, people who would one day emerge as leaders of the toxics movement had only what has been labeled a NIMBY, or Not in My Back Yard, understanding of the problem.

Once engaged and mobilized, many participants moved well beyond NIMBY consciousness. They began to think that the problem was not just one site filled with drums leaking toxic waste. Instead, they began to see the problem as systemic, as the inevitable consequence of how the economy functioned. Getting the responsible corporation to clean up this community or stopping that proposed landfill no longer seemed enough. Real solutions, they began to believe and say, required systemic change. As their understanding grew more sophisticated, they saw their original, local issue in a new light, as part of a larger fight for a safe environment. They sought to form coalitions between their local protest group and other environmental and other social movement organizations. Not every person who was active in a local hazardous waste struggle went through this sort of radicalization, of course, but they were exposed to environmentalist ideas, gained sympathy for those ideas, and, I would say, certainly grew more supportive of pro-environmental government policy.

With this in mind, I return to the political impact of inverted quarantine. When people believe they can deal with an environmental health threat simply by barricading themselves individually with acts of consumption, rather than by trying to confront the threat through activism, the process of politicization never starts. Instead of a threat leading to engagement, it leads to withdrawal into the apparent safety of a carefully cleansed and tended personal space. Summed up over tens of millions of individuals, such behavior not only undermines society's ability and willingness to deal with any one specific issue, for example, willingness to deal with the problems of the public water supply, but it also removes from the stage large numbers of citizens (and again, these will tend to be affluent and generally politically more effective citizens) who, had they not opted for withdrawing into personal refuges, might have become the effective mass base for substantial environmental reforms. Absent mass support for reform, the processes that release large amounts of toxics into the environment continue to do so, unchecked.

Political Anesthesia vis-à-vis the Impending Environmental Crisis

Toxic production and consumption is only one facet of something bigger. With few exceptions, the world's scientists warn that we have mismanaged our relationship with nature and that we are on the verge of triggering an environmental crisis unprecedented in scale and destructiveness.

The public's response? In congressional testimony in 2004, Richard

Clarke, the former White House counterterrorist expert, memorably described CIA Director George Tenet as running around with his "hair on fire" the weeks before the terrorist attack on the World Trade Center's twin towers on September 11, 2001. One can borrow that phrase and say that scientists have been running around with their hair on fire, sounding not just concerned but downright panic-stricken about where things are headed. In contrast, much of the American public seems complacent, unperturbed. According to opinion surveys, the public is aware of the warnings, agrees that the threat is real, but cares about other things much more. It is very much the same pattern that we see in public opinion toward cleaner air and cleaner water: substantial "concern," modest salience.

I believe that mass faith in and use of inverted quarantine products plays a role here too.

Consuming the Planet

Human beings have always actively transformed landscapes and exploited natural resources to satisfy needs. Societies metabolize nature. Materials are "eaten," inhaled; waste products "excreted," exhaled. We cut forests, dam rivers, drain wetlands, graze grasslands, plow up prairie, mine minerals, coal, and metals, and pump oil and gas from below. The by-products of extraction and production, and the stuff we have used and no longer need or want, we throw back.

In the modern age, under the dual impact of staggering economic growth and a vastly increased human population, appropriations from nature and the return of waste materials to nature have reached unprecedented levels. Innovation, growth, and material affluence are the hallmarks of modern society. The environmental consequences, we now know, are unprecedented too. Growth in population, economic activity, resource extraction, and discharges of waste products exploded in the twentieth century.

In the book *Something New under the Sun,* the historian John McNeill presents statistics that together capture the startling pace of development during the twentieth century. Population increased fourfold, from 1.6 to 6 billion persons. Economies grew even faster. World gross domestic product increased fourteenfold; industrial output fortyfold. Ever more resources were being taken from nature. World energy use rose from 1,900 "millions metric tons of oil equivalent" in 1900 to 30,000 in 1990, a fifteenfold increase. In 1900, the people of the world were appropriating about 580 cubic kilometers of freshwater per year; by 2000 freshwater use had grown to 5,190 cubic kilometers a year, a ninefold increase. The amount of land used to grow crops increased from 8 million square kilometers to 15.2 million; land for pasture increased from 14 million square kilometers to 34 million. Ever more waste products were being returned to nature. Carbon dioxide emissions rose by a factor of 17. Other emissions into air grew too: various metals by factors of 5 to 7; methane by a factor of 3.5; sulfur dioxide by a factor of 5; nitrogen oxides by a factor of 14.[24]

Other studies show that our species is now appropriating for our use significant fractions of some of the most basic of the world's resources, water and land. We are taking about half the water in the world's rivers and pumping unsustainable amounts from the world's aquifers. If one excludes uninhabitable land surface (rock, ice, barren land), humans now inhabit and use 36.3 percent of the earth's total habitable land surface. In addition, human activity has "partially disturbed" another 36.7 percent. Only 27 percent remains undisturbed. Twelve percent of the world's usable land surface, 18 million square kilometers, an area about the size of South America, is used for agriculture. By one estimate, humans now appropriate about a third of the earth's terrestrial net primary production; that is, a third of *all* the solar energy that is transformed into plant product through the process of photosynthesis is now either consumed (as food and fiber), or co-opted, or forgone (destroyed) by human activity.[25]

And it still is not enough, in part because the benefits of all this growth have been distributed very inequitably, both between nations and within nations. Billions the world over live in poverty. Hundreds of millions do not get enough to eat. About a third of the world's peoples, 1.7 billion, live in areas that are "water-stressed."[26] People who are a bit better off, such as the peoples of China and India, want more too. They aspire to the standard of living enjoyed by the middle classes in the advanced, industrialized nations.

Furthermore, population is expected to grow to something close to 9 billion before it stabilizes. So, in spite of all the growth over the past century in the consumption of natural resources and in discharges of wastes, pressures for more growth are immense.

Ignoring Dire Warnings

In *The Limits to Growth*, published in 1972, a group of scientists attempted for the first time a computer simulation that would model the interactions among several social and environmental trends on a global scale. The authors' conclusion was that "if the present growth trends in world population, industrialization, pollution, food production, and resource depletion continue unchanged, the limits to growth on this planet will be reached sometime within the next one hundred years. The most probable result will be a rather sudden and uncontrollable decline in both population and industrial capacity."[27]

Although some experts dismissed the Club of Rome's report, arguing that its methodology was flawed, hence its conclusions unwarranted and unduly alarmist, the report was widely discussed, with considerable hand-wringing. If one looks at the numbers, though, one would have to say the report had no real impact. The juggernaut of growth, growth, and more growth just kept rolling on. Between 1972 and 2006, the world's population rose from 3.8 to 6.5 billion. World production of crude oil rose from 51 million barrels a day to 74 million barrels. The U.S. gross domestic product almost tripled, from about $4 trillion to over $11 trillion.

The warnings have continued. They come with more frequency and often with a greater sense of urgency. Consider just one noteworthy example, "Population Growth, Resource Consumption and a Sustainable World," a statement jointly authored in 1992 by the presidents of the Royal Society of London and the (American) National Academy of Sciences:

> The present patterns of human activity accentuated by population growth should make even those most optimistic about future scientific progress pause and reconsider the wisdom of ignoring these threats to our planet. Unrestrained resource consumption for energy production and other uses, especially if the developing world strives to achieve living standards based on the same levels of consumption as the developed world, would lead to catastrophic outcomes for the global environment. . . . Some of the environmental changes may produce irreversible damage to the earth's capacity to sustain life. . . . The future of our planet is in the balance.[28]

Climate Change

Recently, talk of environmental crisis has focused on how human activity is changing the earth's climate. Scientists have reached a "robust consensus"[29] that climate change is real, that it is due to human influence, not just to natural variation in the earth's weather, that we are already seeing the "signals" of climate change today, and that these changes will grow more intense, with dire consequences if the causes of climate change are not addressed soon. That scientific consensus is summarized every five years by the Intergovernmental Panel on Climate Change (IPCC), an international collaborative effort sponsored by the World Meteorological Organization and the United Nations Environmental Programme. The IPCC published its most recent set of reports in 2001.[30] Since then, a veritable flood of new work, too voluminous to cite here, has further confirmed the key elements of that consensus.

The physical evidence is by now incontrovertible. The concentration in the atmosphere of carbon dioxide, the principal "greenhouse" gas, is a third higher than in pre-industrial times and continues to rise each year. The increase in global average air temperature in the past hundred years "is likely to have been the largest of any century during the past 1,000 years."[31] The oceans are warming. Extreme weather events are happening more often, including more frequent and hotter heat waves, record rainfall in some places and prolonged drought in others, and more storms and more intense storms. Glaciers the world over are retreating. Temperatures in both the Arctic and the Antarctic are rising faster than the global average. Sea ice in the Arctic is thinning and shrinking. Permafrost, the frozen soil that covers much of the land mass in the Arctic, is thawing. The glaciers that flow off of Greenland's massive ice sheets are accelerating. Antarctica's ice sheets are shrinking; large pieces of the Larsen Ice Shelf have broken off and drifted away.[32]

Climate change is affecting wildlife. Species' ranges have shifted. The timing of species' reproductive cycles are changing. Many species are said to already be threatened with extinction due to loss of habitat, fragmentation of habitat, poaching, and so on. Climate change, it is predicted, will dramatically increase the rate of species loss.[33]

Likely Social Impacts

Climate models predict that over the next one hundred years temperatures will rise another 1.4 to 5.8 degrees Celsius, most likely between 2.5 and 4.5 degrees. This is "much larger than the observed changes during the twentieth century and is very likely to be without precedent during at least the last 10,000 years, based on paleoclimate data."[34] Some computer simulations suggest the increase could be as high as 11 degrees Celsius.[35]

The climatological and ecological impacts already observed today will continue and will worsen. I am as interested, though, in the likely effects on human societies as I am in the physical and ecological impacts. Reading the literature on the potential social impacts of climate change, I find two distinct scenarios discussed, neither of which are particularly pleasant to contemplate. We can, for convenience, label them the *gradualist* scenario and the *abrupt climate change* scenario.

The gradualist scenario would see food supplies decrease and water grow more scarce just when population growth will increase the need for more of both. Societies will have serious problems meeting basic human needs.

The impact on agriculture will depend, according to the IPCC, on how much temperatures rise. If they rise only another degree or two, agricultural production in the temperate zones of the world could rise, but production would fall "in developing countries— where 790 million people are estimated to be undernourished at present. . . . Climate change would increase the number of undernourished people in the developing world."[36] If the temperature increases are toward the higher end of the estimates, three degrees Celsius or more, the IPCC says agricultural production would suffer the world over.

Access to adequate supplies of clean water is already a serious problem in many parts of the world. "Approximately 1.7 billion people, one-third of the world's population, presently live in countries that are water-stressed." Demand for water for agriculture, industrial production, and personal use can only go up as another 3 billion people are added to the world's population. Even as demand grows, "climate change could further decrease stream flow and groundwater recharge in many water-stressed countries." Communities that depend on snowmelt or glacial melt for their water, for example, could see drastic declines in supply. The numbers of the water-stressed could "increase to about 5 billion by 2025."[37]

Gradual climate change is likely to exacerbate existing inequalities among nations. The IPCC's reports say the affluent societies of North America and Europe have the administrative infrastructures and the collective resources that will be needed by a society trying to cope with the range of challenges posed by climate change. Developing countries are more vulnerable to climate change than developed countries . . . because of their physical exposure to climate change hazards . . . and [because] their adaptive capacity is low due to low levels of human, financial, and natural resources, as well as limited institutional and technological capability.[38]

Differences between nations in their ability to cope with climate change could trigger mass migrations. A Christian relief organization wonders if there will be waves of "water refugees." Others speak of "climate refugees" and of "environmental refugees."[39] If the numbers are large, the nations that the refugees try to enter would undoubtedly attempt to close their borders, perhaps resort to increasingly draconian measures to repel what they would perceive as an unwelcome, destabilizing flood of desperate peoples. On March 22, 2002, World Water Day, U.N. Secretary General Kofi Annan spoke of "fears that water issues contain the seeds of violent conflict."[40]

Now let's look at the *abrupt climate change* scenario. In the early years of climate science, climate change was thought to be slow, incremental. As the field of study matured and data accumulated, that gradualist paradigm had to be abandoned, eventually to be replaced by a new consensus that says that climate is nonlinear, that climate can change abruptly and has, in fact, gone through a number of abrupt, essentially nonlinear shifts before.[41] A National Research Council report, "Abrupt Climate Change: Inevitable Surprises," cites numerous examples of abrupt climate change in the past 100,000 years.[42] Given the right conditions, the report shows, even a small event "may force the climate system across a threshold and trigger huge change."[43]

In its last cycle of reports, the IPCC noted that some climate models show abrupt, potentially catastrophic climate shifts at the upper end of the range of possible temperature increases:

Several climate model simulations show shutdown of the North Atlantic thermohaline circulation with high warming. . . . If this were to occur, it could lead to a rapid regional climate change in the North Atlantic region, with major societal and ecosystem impacts.[44]

The "thermohaline circulation" is an ocean current that carries heat from the equator north, warming Europe and eastern North

America; once the water releases heat, it drops toward the ocean floor and flows back south. Climate scientists believe that when large amounts of fresh (not salty) water enter the North Atlantic, this circulation is disrupted and experiences partial or complete shutdown.[45] Paleoclimate research shows that when the thermohaline circulation shuts down, the climate in Europe (and to a lesser degree in eastern North America) shifts rapidly from moderate and moist to cold and dry. When we think of climate change, we think "global warming" and imagine hotter days and wilder weather, but we should also think, paradoxically, that warming can turn the climate cold and dry in some of the most densely populated regions of the Northern Hemisphere.

The IPCC report tries to be reassuring, noting that "the probabilities of triggering such events are poorly understood" and that "complete shutdown may take several centuries to occur."[46] Recent measurements show, however, that Arctic ice melt has already added large volumes of freshwater to the North Atlantic, and late in 2005 a team of researchers reported that the thermohaline circulation had weakened by 30 percent in the past fifty years.[47]

Abrupt climate change would threaten civilization. Food and water supplies would be threatened even more than they would be in the gradualist scenario. Abrupt climate change would trigger abrupt *ecosystem* changes. Some ecosystems could collapse.[48] Since societies depend so thoroughly and completely on the rich variety of (mostly unacknowledged) "services" ecosystems provide, ecosystem collapse, and even just significant ecosystem change short of actual collapse, could fundamentally compromise societies' ability to "make a living" off nature. The National Research Council's report on abrupt climate change points out that "the collapse of some ancient civilizations has been associated with abrupt climate changes."[49]

And it is not just scientists who are worried. In 2003, the Central Intelligence Agency (CIA) requested an analysis of the national security implications of abrupt climate change. The authors analyze a scenario in which the disorganization of the thermohaline circulation sets off a cold period similar to what is known as the "8,200 year event." Temperatures drop about five degrees Celsius; the cold spell lasts about one hundred years. That scenario is a conservative one. The authors could have chosen to model the impacts of a more drastic change, such as the thousand-year drop of ten degrees Celsius known as the Younger Dryas. Even making the conservative assumption that the drop will not be that severe and the cold will not last so long, the authors write that "modern civilization has never experienced weather conditions as persistently disruptive as the ones outlined in this scenario." Crop yields fall 10 to 15 percent. There are "catastrophic shortages of water." Such conditions "pose a severe risk to political, economic and social stability." The world becomes "increasingly disorderly and potentially violent."[50]

> Humans fight when they outstrip the carrying capacity of their natural environment. Every time there is a choice between starving and raiding, humans raid. . . . if carrying capacities everywhere were suddenly lowered drastically by abrupt climate change, humanity would revert to its [earlier] norm of constant battles for diminishing resources. . . . Once again warfare would define human life.[51]

Scientists' Opinion and Public Opinion

These scenarios describe a world far different than the world we live in today. It is not just a world with no polar bears, not just the end of some societies in the Arctic or Pacific coral atolls, societies whose ecological bases will be completely transformed or will disappear under a rising ocean, as tragic as those consequences might be. No one would be left untouched. The cares and troubles that preoccupy, say, the middle classes in the United States today will seem negligible compared to the scale of disruption of everyday life as the environment grows increasingly chaotic and inhospitable, as our ability to extract a living from nature grows increasingly problematic.

The scientists' hair is on fire. In June 2005, the national academies of eleven leading nations (United States, Japan, United Kingdom, Germany, France, Italy, Russia, Canada, China, India, and Brazil) issued a joint statement, timed to correspond to a summit meeting of the G8 nations, warning of global catastrophe, urging "all nations . . . to take prompt action to reduce . . . net global greenhouse gas emissions."[52]

How is the American public responding to news of these impending developments? Until quite recently, as late as 2003, most Americans did not know much about climate change.[53] By 2006, following the destructive hurricane season of the previous year, years of drought and a summer of heat waves, a burst of discussion in the media, and former vice president Al Gore's film, *An Inconvenient Truth,* public opinion had clearly evolved. Most citizens said they had heard of "climate change" or "global warming." Most agreed that it was happening, and about half thought it was caused by human activity, not just natural variation in the climate system. A majority agreed that climate change is a serious problem, and they said they wanted the government to do something about it.[54]

On the other hand, the polls showed that Americans ranked "global warming" or "climate change" far down the list of issues they considered most pressing, most in need of immediate attention and action. In one poll, climate change ranked thirteenth in importance for Democrats and Independents, nineteenth in importance for Republicans, well behind issues such as terrorism, Iraq, the economy, health care, education, and taxes.[55] Americans said they were not particularly willing to make personal sacrifices that would slow the pace of climate change. When asked if they would be willing to pay more taxes on electricity use or higher taxes on gasoline (such taxes would lower demand for energy and would reduce gas consumption, hence decrease the amounts of carbon dioxide being released into the environment), the response, by substantial majorities, was a resounding no.[56]

Inverted Quarantine, Anesthesia, and Climate Change

Why is it that Americans recognize the reality of climate change but do not put a high priority on politicians addressing it and are not willing to personally sacrifice to help reduce the threat? What explains this kind of mental splitting, or compartmentalization, knowing on some level that the situation is serious but somehow not managing to make it feel *real*?

Some of it is surely due to the fact that the impending crisis predicted by the scientists is at this point somewhat of an abstraction. The social impacts predicted are terrifying, but they describe a world radically different than the world we live in today. It is not the world as we currently experience it, so naturally it does not quite feel real. In March 2006, the Gallup Poll asked, "Do you think that global warming will pose a serious threat to you or your way of life in your lifetime?" Sixty-two percent said "no"; only 35 percent said "yes."[57] Severe climate events, a hurricane destroying New Orleans, hundreds dying in a heat wave can temporarily increase the public's agreement that climate change is real and is serious. The abstract threat has, for a moment, acquired an experiential referent. Then, everything else being equal, that awareness fades.

So, yes, I think that is the main reason why the average American can acknowledge that warnings of imminent environmental crisis are probably right, then go on as if it were not so. The gap between predictions of future catastrophe and people's actual experience—that things are evidently not falling apart now—is simply too great. Still, I would argue that faith in inverted quarantine products also plays a role. Today, most Americans experience the impending environmental crisis, if they experience it at all, in the form of increasing numbers of toxic threats to their health and to the health of their loved ones. Distress about those toxic threats could be the experiential basis from which a sustained, *salient* concern about impending environmental crisis, climate change, and so on could grow. The availability of the inverted quarantine option, no matter how illusory the protection it offers, eliminates this starting point for the development of such concern. Inverted quarantine reduces the perceived, subjective urgency of the situation. When one feels insulated from trouble exactly at the point where the situation would most threaten one's sense of well-being, talk of "crisis" continues to feel abstract, unreal, and far away. People may hear that there is trouble brewing, but when they believe they have taken measures to protect themselves from the most immediate way it seems to affect them, it is harder to really care.

The Shielded Body in an Increasingly Distressed World

Unlike atomic war, the ecological crisis is not going to arrive on one particular day, at one specific hour. The crisis threshold, the point where the routines of everyday life begin to break down, will be approached and then crossed in a more gradual manner, perhaps only to be seen clearly in retrospect. Nonetheless, if nothing interferes with current trends, such a "moment" will come, even if when it does, it arrives over months or years, earlier in some parts of the world than in others, earlier in some segments of American society than in others. Some would say that moment is already here in various places around the world.

At that moment, it will become painfully clear, even to its most ardent practitioners, that inverted quarantine will not shelter anyone, even the most privileged person, from what is happening. The fundamental flaw in the logic of the inverted quarantine was identified early on by the critics of the family fallout shelter. They pointed out that even if some individuals came through the bombardment and emerged from their shelters, they would open their door to a world in which social organization had broken down, weather had changed, farmland had been poisoned, industry destroyed, ecological systems disrupted, a world that could no longer sustain human social life.

The shelter critics' insight into the futility of individuals physically surviving for a couple of days or weeks following an all-out exchange of hydrogen bombs went right to the heart of what is most wrong with the inverted quarantine response to any hazard that is not just some isolated, self-contained problem but a manifestation of bigger things going wrong. Inverted quarantine is implicitly based on denial of complexity and interdependence. It mistakenly reduces the question of an individual's well-being to nothing more than the maintenance of the integrity of the individual's body. But in the modern world a person's well-being cannot be reduced to preservation of the person's body alone. Each person's well-being requires in our age that somewhere else in the world innumerable other people grow that person's food, manufacture the goods he uses, and help make all the organizations he depends on continue to function day to day. All that, in turn, requires that human societies engage in successful, more or less stable exchanges with the natural environment.

It is exactly that sustaining exchange with the environment that will break down if conditions begin to resemble some of the more dire scenarios described by scientists. What good will individual acts of inverted quarantine do then? What good does it do to encase one's body in a toxics-free bubble when things are falling apart so completely? Drink bottled water? Sure, why not. Eat organic, buy natural clothing and furniture—perfect your bubble. Didn't people once build fallout shelters in the hope that they would get them through a full-scale nuclear war? In the long run, inverted quarantine in response to environmental threat could be just as futile and illusory as that backyard fallout shelter.

If the political anesthesia, the complacency, inverted quarantine fosters does contribute to our collective inability to stop the world from rushing on toward crisis, one can say that in the long run inverted quarantine will help create the conditions of its own complete and utter failure.

Conclusion:
The Future
of an Illusion

Books about environmental crisis tend to follow the same general narrative arc: First comes bad news and dire warnings. "We face catastrophe if we continue down this path." Then, when the author gets toward the end of the book, she or he affirms that there is still time. Take heart. It is not too late. There is still a chance that we can turn things around, if only we change our ways: If we embrace a new, more egalitarian, more humane, non-exploitative attitude toward nature. If we can find our way to a new economics that creates acceptably high levels of material well-being without that relying on ceaseless exponential growth. Or, at least, if we dramatically reduce our reliance on oil and switch to renewable, nonpolluting sources of energy, reduce the amounts of toxics we release back into the environment, and so on.

It makes good psychological sense to hold out the possibility that we can still change things for the better, that it is not too late. You have to make people look bad news in the face. You have to disturb them, scare them. But if you leave it at that, perhaps all you will manage to do is foster a sense of hopelessness and resignation: "Change all that? It's overwhelming to contemplate. Impossible. It's too late. There's nothing to be done. I might as well phase back into denial, soldier on, worry about the things I *can* do something about." If the point is to alarm people so that they will then act—do *something*—you have to affirm that it is still *possible,* that the game is not over, that there is still time.

And there is a certain deep truth in that. The future is the place where hope finds a home. The future is not completely open; it is not just pure possibility. Still, the future is open to a range of possibilities, so no matter how bad the situation seems right now, we still have the opportunity to fix things.

I wish I could follow that recipe and simply conclude that although the situation looks bleak, there is still hope, as long as we begin to make those changes now, or at least soon, before we cross some sort of threshold or tipping point. To be completely frank, though, I can repeat the formula only if I add a certain caveat. I *can* say that good things would happen if everyone refused the siren call of inverted quarantine en masse and began, instead, to insist on real solutions to our very real social and ecological problems. I am forced, though, to add, however regretfully, that as far as I can see, conditions do not favor such a refusal. We are much more likely to see people continue to flee to inverted quarantine ever more desperately for the foreseeable future.

Good Things Happen When People Refuse Inverted Quarantine

If we examine those instances when society refused the mass inverted quarantine option, we quickly come to see that that refusal opens the way to real solutions, to substantial reforms that leave conditions far better than they would have been otherwise. Let's consider some examples.

In the case of the fallout shelter panic of 1961, the critics warned that mass shelter building would only increase the likelihood of war. At the height of the panic, about half of all Americans said they were considering building a backyard or basement shelter. Then panic subsided. Only a few hundred thousand families built home shelters. The vast majority of citizens concluded that the antiwar critics were right. A fallout shelter would be worthless.

In a sense, this was an easy victory for the anti–inverted quarantine side, but that should not detract from the significance of the victory. Having seen pictures of Hiroshima, having repeatedly seen footage of the suburban dream house being blown down by an atomic test in the Nevada desert, the remains bursting spontaneously into flames, most people ultimately could not make themselves believe that the atomic bomb is just a bigger bomb, that a rain of atomic bombs did not necessarily mean the End of the World, that they could "think about the unthinkable," that they would emerge from those backyard shelters, dust themselves off, roll up their sleeves, and get to work rebuilding their lives.

The fallout shelter case shows how rejection of the inverted quarantine alternative can open the path to more constructive answers. Rather than two peoples digging holes in the false hope of surviving the nuclear holocaust, after 1961 the United States and the USSR began to slowly work their way toward detente. Mutual interest in backing away from atomic war was severely tested by the Cuban missile crisis in 1962, a harrowing incident that temporarily raised tensions between the superpowers nearly to the breaking point, but the general direction of bilateral relations—away from atomic confrontation—prevailed. Kennedy and Khrushchev signed the first test ban treaty in 1963. Kennedy told the nation that the test ban meant "escape from the darkening prospect of mass destruction. . . . a step toward peace—a step toward reason—a step away from war."[1] That first ban on atmospheric testing was followed by the nonproliferation treaty (1970), antiballistic missile treaties (1972, 1974), the underground test ban (1974), and SALT II's limits on strategic offensive weapons.

I am not arguing that there is a simple cause-and-effect relationship here, that rejection of inverted quarantine automatically and necessarily resulted in a better outcome. I am suggesting instead something more conditional and open-ended. Rejection of inverted quarantine was akin to refusing to make a wrong turn that would only have led to a dead end. Not taking that turn did not guarantee anything. "All" it did was to keep open the possibility that better choices could yet be made. But that is a lot. Better choices *were* made. Had millions of Americans gone ahead and built atomic shelters, that would have heightened tensions and increased the likelihood of war breaking out. When that did not happen, when Americans decided there was no point in pretending an underground room and two weeks of supplies would help them survive such a war, the nation's leaders pulled back from the brink and began instead to forge, slowly and cautiously, an increasingly effective set of international agreements that, together, increased trust, decreased the likelihood of misunderstanding, and substituted diplomacy for posturing.

Let's consider a more mundane case, some sort of toxic environmental threat. Let's say that potential victims, scientists, public health activists, the media and other opinion makers, and elected officials or experts in a regulatory agency decide that some substance poses a threat. How can individuals and society as a whole respond?

One way to respond is for those who can afford it to try to buy themselves protection, individually. They purchase and use products that claim to keep that substance from entering and thus polluting their bodies. I have discussed the problems with this approach. Environmental inverted quarantine products are often ineffective, but because people believe—falsely—that they are protected, they are less likely to feel an urge to voice support for the kind of regulatory controls that would be needed to really address the hazard. Insufficient support translates into insufficient incentive for regulators. Since the substance is not regulated, or the regulations meant to deal with it are not stringently implemented, the substance continues to circulate unabated in the environment.

Far better, for both individual citizens and for society as a whole, is for the public to shun the inverted quarantine option and back instead substantial regulation of the production processes or of the waste disposal practices that generate the hazard in the first place. Americans benefited substantially from enforcement of the many standards promulgated under the authority of the Clean Air Act, the Clean Water Act, the Toxic Substances Control Act, the Federal Insecticide, Fungicide, and Rodenticide Act (FIFRA), and the Resource Conservation and Recovery Act (RCRA, the law that regulates disposal of toxic industrial wastes). They would have benefited more had these laws been more stringently interpreted and enforced, and even more had Congress put more teeth in these laws in the first place.

Data from the Centers for Disease Control's biomonitoring studies, the Human Exposure to Environmental Chemicals studies, show the benefits of strong regulation in the most direct and powerful way imaginable.[2] The studies show that when a substance has been completely banned (DDT, PCBs), or when it has been very rigorously regulated (lead), or when a successful public health campaign has reduced its consumption (cotinine, a metabolite of the nicotine found in cigarette smoke), concentrations of that substance in people's bodies, in their blood or in their urine, goes down over time. The evidence is clear. If a substance is circulating in the environment, inverted quarantine measures might or might not—often not, as we have seen—keep that substance from entering the body. If that substance is not in the environment in the first place, what or how one breathes, drinks, or eats is irrelevant. No extraneous substance out there to ingest? No worry and no need to do anything special to keep it out of the body.

The very best course of action is what environmentalists call the precautionary principle. New products are not made and new ways of making existing products are not put into operation until it has been shown that they do not unduly harm either the environment or public health, that there are no serious side effects at any point in the product life cycle from the point that they are made to the point that they are used up and discarded. Short of that, given that we are not ready yet as a society to commit ourselves to the precautionary principle, the second best option is strong regulation, especially regulation that bans the production and prevents the circulation of hazardous substances, rather than merely regulating how that substance is managed once it has been created.

A Recent Example: Different Ways to Deal with the Ozone "Hole"

Sometime during the 1970s, scientists from the British Antarctic Survey noted that levels of stratospheric ozone above the South Pole had dropped. (The survey team would not publish its findings until 1985.) The phenomenon was soon dubbed the "ozone hole." Since then, the "hole" has reappeared annually, growing with the onset of the Antarctic winter, "closing" in the spring (i.e., ozone levels recover), only to return the next winter, generally larger and the level of ozone depletion within the "hole" more severe than the year before. In 2005, the Antarctic hole was reported as "one of the deepest and largest recorded . . . with ozone amounts over 50 percent down on the normal for the time of year in many places . . . [the hole reaching a] peak of 25 million square kilometres." A similar, though smaller, hole was subsequently detected over the North Pole. Stratospheric ozone has thinned over other parts of the globe too, though less so. "In the middle latitudes (most of the populated world), ozone levels have fallen about 10 percent during the winter and 5 percent in the summer."[3]

The main substances responsible for atmospheric ozone loss are chlorofluorocarbons (CFCs), first synthesized in 1928, produced since 1930 in very large quantities, used most commonly as a refrigerant in refrigerators and air conditioners.[4] (Other substances, halocarbon solvents, methyl bromide, a fungicide used in agriculture—in my community it is applied to fields before a strawberry crop is planted—also contribute to ozone depletion.) When molecules of CFC escape into the open environment, they begin a long, slow journey up through the atmosphere. Once they reach the stratosphere, solar UV radiation breaks them down, releasing free chlorine atoms. The chlorine atoms interact with ozone (O_3) molecules, converting them into molecules of ordinary oxygen (O_2). NASA's ozone history Web page reports that "one chlorine atom can convert 100,000 molecules of ozone" before that chlorine atom combines with some other atom to form a molecule (e.g., HCl) that no longer attacks and breaks down ozone molecules.[5]

For decades, but unbeknownst to anyone, a common substance found in millions of household appliances had been eating away at atmospheric ozone. The amount of ozone in the upper atmosphere matters because ozone blocks solar ultraviolet radiation. If the ozone layer thins, more UV light reaches the surface of the earth. Excessive exposure to solar radiation in the UV range has been shown to increase rates of skin cancer and eye diseases, such as cataracts. UV can damage the immune system.[6]

If the causes of ozone loss were not dealt with, the "hole" would eventually expand to the point where substantial human settlements, at the southern tips of Africa and South America, eventually in Australia and New Zealand, would be "under" the hole. Many people would then be exposed to very high levels of solar UV. Recall, too, that ozone levels had already fallen, though less so, in the temperate latitudes—over the most heavily populated parts of the planet. If ozone levels continued to fall worldwide, billions of people would be at risk.

Not just people would be affected. Right after the ozone hole was discovered, quite a lot of concern, one might even say alarm, was expressed about impacts on other species, on whole ecosystems. Much of that concern focused, naturally enough, on Antarctica and the oceans around it. A lot of the discussion was about the potential effects on phytoplankton living near the surface of the oceans, because "phytoplankton are the most important biomass producers in aquatic ecosystems."[7] A significant decrease in biomass productivity at the bottom of the oceanic food chain would have a major impact on the whole ecology of those oceans. As ozone levels fell over other parts of the planet, ecological systems elsewhere could be affected too, it was said.

What would the world's nations do about the ozone hole?

Many early articles about the ozone hole and why it should concern us focused on people's health, especially on the links between ozone loss, higher levels of UV flux, and predictions that skin cancer rates would skyrocket. The response could easily have been that people should apply more sunscreen, wear sunglasses that block out UV rays, and wear hats. Such a "take" on the problem would have dovetailed well with the anti-tanning campaign already under way at the time. Dermatologists and public health activists had campaigned for years against the widespread practice of suntanning. They urged people to avoid too much sun exposure, to wear long-sleeved shirts and hats, to stay out of the sun during peak hours in the middle of the day, and if they did go out, to apply a sunscreen that offered UV protection. Since the anti-tanning public health campaign had been going for some years and was well established, it would have been easy for the response to have taken the form of an intensification of that campaign, to simply tell people to do even better avoiding sun exposure, be even more diligent applying high-SPF sunscreen, and so on.[8]

A campaign to get people to use sunscreen more without any further societal action to actually reduce the emission of ozone-depleting substances would have been a classic, textbook example of the inverted quarantine response. In this instance, though, the inverted quarantine option was shunned, and the world's governments chose instead to deal with the problem head on. Starting with the Montreal Protocol (1987), signatory nations agreed to start phasing out production and use of CFCs. Later revisions of the protocol strengthened controls over CFCs and added controls over other ozone-depleting substances.

We can consider what *would* have happened if no international treaty had been negotiated and the world's peoples were simply advised to make greater efforts to protect themselves. Doing so will give us a sense of how beneficial rejection of the inverted quarantine "solution" was in this case.

First, since no actual limits would have been placed on ever more CFCs and other ozone-depleting substances being released into the atmosphere, ozone depletion would have continued to worsen.

People would have been told to be ever more vigilant, to modify their activities so they would be out of the sun in the middle of the day, or to wear protective clothing (long-sleeved shirts, hats) if they went out. They would have been admonished to always use sunscreen.

Would that have worked?

At the very least, that response would have required considerable sacrifice in terms of losing what one might think of as the freedom to move around without worry or constraint. The first line of defense, staying indoors, is really just another form of self-incarceration (see chapter 2), a voluntary narrowing of one's freedom of movement, this time not for fear of the Dangerous Classes but for the sake of limiting one's exposure to sunlight. That is an incalculable existential price to pay, living with the ever-present fear that sunshine can kill you.

Sunscreen has its problems too. Sunscreen products are not cheap, especially when applied as often and in the amounts recommended by the manufacturers.[9] Few are able to spend what it would take to use sunscreen religiously, day in and day out, over a lifetime, every time they went out, so income is once again a factor in who can avail themselves of this "solution" and who cannot. On the other hand, there is some evidence that sunscreen may not actually be effective in preventing skin cancer, even at current levels of UV flux.[10] If so, even those who can afford to use sunscreen in copious amounts might not have been able to protect themselves from the harm inflicted by ozone loss.

At the same time, if nothing substantive had been done to stem and reverse the loss of ozone, the impact on various ecosystems could have been quite serious. Given the current state of knowledge, it is not clear how ecosystems would have fared. On the one hand, the catastrophic impacts some predicted soon after the ozone hole was discovered have not happened. Plankton, some plants, and marine life have shown more ability to adapt to higher UV than first predicted.[11] Scott Norris quotes Deneb Karentz, a marine biologists who has studied ozone and UV effects in the Antarctic region, saying, "The Antarctic ecosystem is not going to collapse because of the recent ozone depletion."[12]

On the other hand, significant impacts *have* been observed. Phytoplankton, macroalgae, and sea grasses, which are also "important biomass producers,"[13] and zooplankton communities have all been adversely affected.[14] "Significant changes of solar UV radiation on aquatic ecosystems may result in decreased biomass productivity. The impact of this decrease would be reflected through all levels of the intricate food web, resulting in reduced food production for humans . . . reduced sink capacity for atmospheric carbon dioxide . . . , as well as changes in species composition and ecosystem integrity."[15] As Karentz told Norris, although the "Antarctic ecosystem is not going to collapse . . . it may end up looking different."[16]

M. M. Caldwell and colleagues (1998) reviewed studies of terrestrial ecosystem impacts and found evidence that "ecosystem-level consequences . . . are emerging." They write, "the overall productivity of the system may well remain about the same while species composition may change. However, a change in the balance of species could have far-reaching consequences for the character of many ecosystems."[17]

Researchers emphasize how uncertain they remain:

> After more than a decade of research, much information has been gathered about UV-photobiology in Antarctica; however, a definitive quantitative assessment of the effect of ozone depletion on the Antarctic ecosystem still eludes us. . . . The long-term consequences of possible subtle shifts in species composition and trophic interactions are still uncertain.[18]

> Damage to [aquatic] ecosystems is still uncertain. . . . complex rather than simple responses are likely to be the rule.[19]

> [terrestrial ecosystem effects have] thus far received very little attention in experimental research. . . . ecosystem-level consequences . . . are emerging and their magnitude and direction will not be easily predicted.[20]

What if society had chosen the personal protection, sunscreen, hats-and-long-sleeves response while allowing continued production, use, and release of CFCs and other ozone depleting substances? It would have proven to be another classic case of inverted quarantine: individuals *falsely* believing that they have managed to protect themselves, while collective, ecological conditions are allowed to continue to deteriorate, with uncertain, potentially quite dire consequences.

We are fortunate that the inverted quarantine path was not taken and that instead the world's nations joined together to sign the Montreal Protocol and have agreed to strengthen that original treaty several times since. The ozone hole still appears each year and, given the time lags built into the process, has not begun to shrink yet, but scientists expect that levels of stratospheric ozone will soon begin to recover and will recover completely sometime after the middle of the century.[21]

Conditions Favor Greater Dependence on Inverted Quarantine

We should be greatly encouraged by the fact that the world's nations were able to forge treaties to reduce amounts of ozone-depleting substances. It shows that society is still able to choose the more rational alternative to inverted quarantine. I believe, however, that in the near term conditions favor movement in the other direction, toward greater and greater reliance on inverted quarantine.

Almost a Reflex

Let's first observe that the tendency to respond to a new threat by stampeding to an inverted quarantine "solution" is now so deeply ingrained in the American psyche that it seems almost automatic, akin to a bodily reflex, as opposed to a considered and reasoned choice.

Many people feared society would be brought to its knees by the so-called "Y2K bug," the programming error that was predicted to make computers malfunction when their internal clocks tried to roll over from 1999 to 2000. Anticipating "bread lines, bank runs, power failures and looting . . . social chaos," some folks went into survivalist mode, stocked up on freeze-dried food, bottled water and water purifiers, and medical supplies, and went out and bought a generator.[22]

After the terrorist attack brought down the World Trade Center towers on September 11, 2001, Americans, deeply anxious, rushed to insulate themselves. How they tried to do that—what they bought—depended, as always, on ability to pay.

At the top of the class hierarchy, there was increased demand for private jets, as the elite sought to avoid both the commercial flights that might again be hijacked and also the hassles of having to go through more stringent airport security, elbow to elbow with other air travelers.[23] There was also increased demand for armored luxury cars. Long before 9/11, some people, celebrities, politicians, corporate CEOs, and drug dealers, fearing being attacked or kidnapped and held for ransom, were having their cars armored. For a mere $30,000 to $300,000, a car could be retrofitted with armor and made bulletproof. BMW and Mercedes-Benz were selling factory-armored luxury models, BMW the 760Li High Security and Mercedes an armored version of the S500. After 9/11, demand for armored luxury cars jumped so much that Ford planned to roll out a Lincoln Town Car BPS (BPS for Ballistic Protection Series), and General Motors planned to begin offering an armored version of the Cadillac DeVille.[24]

Even before 9/11, some people who could afford it were having "safe rooms" built in their homes, secure, hidden, "impenetrable . . . space to retreat to in the event of an armed intrusion."[25] After 9/11, one entrepreneur offered for sale a "life cell," which could "transform an ordinary living room into a safe, self-sufficient oasis in the midst of a bioterrorist attack,"[26] as well as a larger unit one can bury in the backyard that will keep a family safe for two or three weeks following an act of "nuclear-biological-chemical terrorism," or even following a "full-scale nuclear-biological-chemical war."[27] Some home owners, fearing nuclear terrorism, started building fallout shelters again.[28]

Those measures are for the home. What about at work?—an obvious concern after the World Trade Center towers went down. Soon after 9/11 and the still unsolved anthrax attack that followed,[29] National Public Radio's Daniel Zwerdling reported that parts of some office buildings were being retrofitted to create in them "safe havens" where the air could be filtered in case there was a bioterrorist attack.[30]

What about those further down the class ladder, who cannot afford to fly by private jet, drive an armored Mercedes, have a safe room at home, and work in a building that can protect its occupants in case of a bioterrorist attack? What actions could they take if they felt vulnerable? After 9/11 there was a run on guns,[31] and also a run on bottled water.[32] Attending the 2004 International Builders' Show in Las Vegas, a journalist reported displays of "all manner of newfangled security devices" for "a nation of households [that] imagines itself under siege. . . . Perimeter cameras, alarms and other deterrents on the outside . . . infrared sensors inside."[33]

Perhaps nothing shows more clearly how much the inverted quarantine response has become something like a reflex than the duct tape panic buying episode of 2003. The Department of Homeland Security raised its color-coded terror threat to *orange,* and Americans rushed out to Wal-Mart, Home Depot, and local hardware stores to stock up on duct tape and plastic sheeting.[34]

Inverted Quarantine as *Mentalité*

To call inverted quarantine a "reflex" is not entirely appropriate, of course. The term does capture some of the qualities of the inverted quarantine response, the way stimulus leads to response without having to be mediated, seemingly, by deliberative thought. But it is not exactly right to liken it to a completely automatic physiological response. Since we are considering a matter not of body but of mind, that is, of some combination of perception, assessment/evaluation, intention, and action, I think the more apt concept here is *mentalité*, a term that comes to us from the French *Annales* school of historiography.[35] Similar to the English expression *mentality*, the term attempts to capture the idea that members of any social group share certain "habitual or characteristic mental attitude[s] that determine how [they] will interpret and respond to situations."[36] Thinking of inverted quarantine as a manifestation of contemporary *mentalité* strikes me as correct, more appropriate than to describe the act as a "reflex," but still capturing the sense that this way of reacting to threat is today very deeply rooted, no longer a conscious *choice*, more a preconscious, semiautomatic, unreflectively self-evident, seemingly "natural" way to respond to things.[37]

I pause to reflect that since inverted quarantine is the core idea here, this book should be considered not just a work in environmental sociology but also as an essay in cultural sociology, a study of contemporary mass psychology and of a certain feature of contemporary political culture.

Deregulation and Privatization

Examples like the run on duct tape make the inverted quarantine reflex seem absurd, just plain nutty. One cannot ignore the fact, though, that certain real-world conditions reinforce an individualized, self-reliant response to threat, making it seem not so unreasonable. For, if you feel at risk, who, other than yourself, can you depend on?

Government regulators? Sure, if effective regulatory laws were in place and those laws were being fully, assiduously implemented. They are not.

The wave of new environmental regulatory laws that were passed in the late 1960s and the 1970s, the Clean Air and Clean Water

Acts, the Toxic Substances Control Act, the Resource Conservation and Recovery Act, and others, left Americans far better protected than before. True, even as originally written these laws were not as powerful as they could have been. Compromises were made as the various pieces of legislation were crafted. For instance, when Congress was discussing what the federal government should do about the problem that every year hundreds of billions of pounds of toxic industrial wastes were being released into the environment, it shied away from simply making industry generate less waste and opted instead for a cumbersome system that would try to regulate how wastes were to be disposed.[38]

The choice Congress made in that case is typical. As regulatory laws are written, legislators strike compromises, trying to find balance between the need to protect public health, on the one hand, and their wish to avoid interfering *too* much with the workings of the private economy, on the other. Legislators are exquisitely aware that the industries they are about to regulate do not like it and that they want to be left alone, free to pursue their business however they want.

Furthermore, regulatory laws are hardly ever implemented as stringently as they should be. All the many social scientific studies that have looked at the question of implementation agree that regulations are tough to implement properly even under the best of circumstances.[39] Those best circumstances rarely, if ever, exist. Regulatory agencies do not get the budgets they would need if they were to properly carry out all the tasks mandated by the statutes. Regulated industries employ scientists, attorneys, and lobbyists who monitor every detail of regulatory policy implementation and quickly mobilize when necessary to lobby, testify, submit their views, submit data supporting their position, if necessary, sue—do whatever it takes to reduce agencies' abilities to regulate their activities.

Implementation can be weakened, finally, when the administration in office is hostile to the mission of the regulatory agencies. Regulatory agencies came under sustained attack during the first Reagan administration, 1981–84. Their situation stabilized, but did not necessarily improve, during the Clinton years. The younger Bush's administration is as hostile to environmental protection as Reagan's was, if not more so. Since taking office in 2001, it has done everything in its power, comprehensively and in detail, to further weaken federal environmental regulations,[40] leaving the public far less protected than before. If I were feeling vulnerable, I would not look to federal regulators to protect me at this point, and I certainly would not expect the government to do more any time in the foreseeable future to better protect me from toxic substances in the environment.

When government regulates, it says to its citizens that government has the responsibility to protect them from harm, that the government accepts that it has the duty to make sure that conditions in civil society—in other words, the collective conditions of social existence—do not threaten citizens' ability to live a good and decent, happy and healthy life. Deregulation signals the opposite philosophy of governing, that government's responsibilities are much more narrow and limited, that government should be very reluctant to interfere with private economic activity even if that activity harms individuals.

Deregulation is only one facet of a larger trend of officeholders turning away from the philosophy of government embodied by the New Deal and Lyndon Johnson's Great Society, that policy should, to some degree, shield vulnerable members of society from the most severe impacts of impersonal, essentially pitiless market forces. The philosophy that has now replaced that earlier one calls not only for deregulation but also for deep cuts in the social "safety net" for the poor and the unemployed. What you do not cut outright, you try to privatize, as with President Bush's 2005 campaign to privatize Social Security. In what is undoubtedly the most revered inaugural address any American president has ever delivered, John Kennedy said, "My fellow Americans, ask not what your country can do for you. Ask what you can do for your country." If the reigning political philosophy were reduced to a similar slogan, it might be: "Don't bother asking what your country can do for you. It's going to do less and less. So you just go and figure out how you're going to take care of yourself."

In the introduction, I said inverted quarantine was a resigned or fatalistic expression of environmentalism. The people who resort to it recognize that there is a problem and are in fact quite distressed by that problem and intent upon doing something about it. Such people, however, are deeply pessimistic about real change, unable to imagine that things can actually improve, and therefore fatalistically resigned to it being a dangerous world. The only course of action left to them, they feel, is to try as best they can to shield themselves individually from harm. To say that such an attitude expresses environmental awareness and concern but does so in "twisted and perverse form" is, obviously, to judge such an approach pretty harshly. On the other hand, if one considers the conditions I have just described, that we live in a world where regulations are weak and unlikely to get better any time soon, and in a political culture that urges citizens *not* to ask what their government can do for them, it is not completely irrational for a person who feels vulnerable or at risk to think *first* to do something, anything, individually, to take care of himself or herself.

Feedback and a Downward Spiral toward a Breaking Point

As one tries to divine what the future holds, the most obvious, most "natural" procedure is to extrapolate from what is happening now. Simple extrapolation has its pitfalls, however. History has its twists and turns, surprising conjunctures, and accidents. Unanticipated events can, in the right circumstances, prove decisive and determine subsequent developments. Mindful of that, can we discern where current trends are likely to take us?

For all the reasons I put forth just now, I believe that in the near term Americans will continue to embrace the inverted quarantine option at least at current levels and probably with even greater intensity.[41] At the same time, economies are still expanding and population still growing, so, barring significant changes, environmental conditions in the United States and globally will continue to deteriorate.

The two trends, worsening conditions and ever more desperate flight to the imaginary refuge of inverted quarantine, are likely to reinforce each other, fueling a positive feedback loop: Absent the collective will necessary to undertake substantial, systemic reforms, nothing, or too little, will be done, and conditions will continue to deteriorate. Worsening conditions, in turn, will spur ever more heroic inverted quarantine efforts.

Giving Up Illusion, the Sooner, the Better

One can imagine this feedback loop spiraling on until a moment when conditions approach the point where the routines of everyday life begin to break down. At that point everyone will see that even the most obsessive application of inverted quarantine strategies will do no good. By then, though, conditions will be far worse than they are today. Many more people will be suffering; recovery will be much more difficult. All of us—other species, the whole planet—would be far better off if we could give up the comforting illusion of individualized self-protection long before we reach such a moment.

Practically speaking, I do not know how exactly we can get there, but I am quite sure that to start moving in the right direction we must give up the fantasy that inverted quarantine can save any of us. Even if we do manage to do that, though, it will not be enough. Just giving up illusion by itself will not solve anything. New ways to produce goods with less energy, cleaner energy, and using fewer toxic materials will have to be found. The whole way we build cities and use land will have to be rethought. We will have to question as never before how we have built our lives around the automobile. And these are only the most obvious items on the long list of things we must question and eventually change. We will be able to undertake that conversation in earnest, though, only after we have thoroughly disabused ourselves of the seductive promise of inverted quarantine and are no longer partially anesthetized by the false sense of security it offers us.

I am not saying that we should stop eating organic food and instead willingly gorge on pesticide residues in solidarity with the world's poisoned masses. Every person has the right to do whatever is necessary to live their life without toxics entering their bodies. All I am saying is that while we continue to strive to keep our and our children's bodies healthy, we must not lose sight of the fact that that is not enough. And I am saying that the critique of inverted quarantine needs to be made one of the key campaign points in the next phase of the environmental movement.

It is heartening to remember that the shelter critics won *their* struggle to convince Americans that building fallout shelters was folly and suicidal. Rationality prevailed over the illusory siren song of individual self-protection. That fight needs to be fought—and won—again, this time concerning environmental threats, not nuclear ones.

Notes

Introduction

1. Beverage Marketing Corporation of New York, "News Release: Bottled Water Continues Tradition of Strong Growth in 2005," April 2006, www.beveragemarketing.com (accessed September 18, 2006).

2. *Beverage Industry/Annual Manual* 1989/90: 47.

3. *Beverage Industry* 2000: NP26. By 2003, consumption of bottled water had caught up with consumption of milk, coffee, and beer—all around twenty-one to twenty-two gallons per capita—and was "poised to surpass its competitors to become the second most popular commercial beverage in the U.S." www.bottledwater.org/public/BottledWater2003.doc.

4. Social movements can take even what most would consider, intuitively, a private or personal misfortune and redefine it as social or collective in origin. An excellent example is the way modern feminism considered women's psychological pain a symptom of patriarchal oppression, insisting that "the personal is political." The grassroots toxics movement did something similar when it told people living in contaminated communities that the miscarriages, birth defects, and cancers they and their families were having were not just private or personal tragedies but were also probably caused by irresponsible corporations that had polluted their environment.

5. Flacks 1988.

6. Some time after I arrived at this term I learned that a similar expression, "reverse quarantine," had started to appear in the public health literature. Reverse quarantine is an emergency public health measure meant to protect the residents of a city or a neighborhood if the area is threatened by a sudden release, accidental or purposeful (as in a terrorist

attack), of a harmful chemical, biological, or even radioactive substance. It is related to the idea of "sheltering in place" in case of an industrial accident. The idea is that in such circumstances the best thing officials can do is order the local population to stay put, close doors and windows, and refrain from going out and moving about, either until the hazardous material dissipates on its own or is gathered up and removed by teams specially trained for that kind of work.

The two ideas, reverse quarantine and inverted quarantine, do have some things in common: they both invert the dyadic opposition upon which the logic of classical quarantine is based—healthy conditions overall/diseased individuals. Nevertheless, the two terms stand for two completely different kinds of activities. Reverse quarantine is a rational public health response to a particular type of threatening condition. The threat occurs at a specific time and place; officials direct the action, in some coordinated manner; when the situation is resolved, the "all clear" is sounded, and life returns to normal. With inverted quarantine, the threat is diffuse, with no clear boundaries in either space or time. It is, we imagine, everywhere, ongoing, chronic, with no end in sight. Perhaps most important, inverted quarantine is not formal public health policy triggered by and rationally designed to respond to a specific type of threat. It is, rather, a spontaneous, uncoordinated effort by many individuals, all trying, by themselves, to protect themselves from threats real or imagined. As I discuss later, in the concluding chapter, inverted quarantine is a manifestation of a certain way of perceiving and reacting, a manifestation of what associates of the French Annales school of historiography call *mentalité*.

7. For Paris, see Chevalier 1973; for London, Schupf 1971 and Symons 1849; for New York City, Brace 1967 [1880].

8. Early, elite suburbanization is described by Fishman 1987 and by Jackson 1985.

1. The Fallout Shelter Panic of 1961

1. J. F. Kennedy 1961.

2. Waskow and Newman 1962: 10; Weart 1987: 22. More generally, the account in the introductory section of chapter 1 is based on Waskow and Newman 1962; Watson 1984; and Weart 1987, 1988.

3. "Fallout Shelters," *Life* 51, 11 (September 15, 1961): 94.

4. Waskow and Newman 1962: 11–12.

5. Weart 1988: 259.

6. McHugh 1961: 824.

7. Waskow and Newman 1962: 88.

8. Kaplan 1983: 312; Schlesinger 1965: 747.

9. Schlesinger 1965: 747.

10. Waskow and Newman 1962: 101–25; Singer 1961; Civil Defense Letter Committee 1961. Similar arguments continued to be made after the panic subsided: Fromm and Maccoby 1962; Inglis 1962; Federation of Atomic Scientists 1962; Feld 1962; Piel 1962.

11. Waskow and Newman 1962: 85.

12. Boyer 1994. The emotional climate in the immediate postwar period is also described by Oakes 1994, especially 43–46; and by Weart 1987, 1988.

13. Quoted by Boyer 1994: 7.

14. Quoted by Boyer 1994: 14.

15. Don Goddard, in a radio newscast the day the White House announced the bombing of Hiroshima. Quoted by Boyer 1994: 4.

16. Boyer (1994) cites a number of references to the bomb as Frankenstein's monster, 5, 9.

17. "The War Ends: Blast of Atomic Bomb Brings Swift Surrender of Japanese," *Life* 19, 8 (August 20, 1945): 25–31.

18. Hersey 1946: 51.

19. Ibid.: 26.

20. Ibid.: 76.

21. Oakes 1994: 45.

22. Cited by Oakes 1994: 44.

23. We can get a good idea of people's fears by looking at the *reassurances* offered by official spokesmen. In *How to Survive an Atomic Bomb* (U.S. Federal Civil Defense Administration 1950), Richard Gerstell, Senior Radiological Safety Monitor for Operation Crossroads (a series of tests at Bikini Atoll) and a consultant for the Federal Civil Defense Office, says that many people have "silly ideas . . . completely wrong notions . . . about the atomic bomb" (1). Apparently people were worried about radioactivity or "atomic rays." Gerstell says, "Yes. The bomb does give off dangerous atomic rays. But many completely wrong statements are being made about these rays. . . . this 'fall-out' stuff . . . can cause burns if you were exposed to it long enough. But it is not likely to hurt you. . . . you can easily protect yourself from it. . . . Not one person in Hiroshima or Nagasaki was killed or injured by lingering radioactivity. That is a *fact*" (14; italics in original). People must also have thought that exposure to "atomic rays" could cause cancer, infertility, and birth defects: Cancer? "*That is absolutely false.* . . . many thousands of [Japanese] people were exposed to atomic rays. . . . Not one ray-caused cancer in all those thousands of cases" (15; italics in original). Infertility? "If you are properly shielded from the rays . . . you will not become unable to have children" (16). Birth defects? "The rays will not necessarily make you have children who are freaks. Not one of over 12,000 carefully watched Japanese survivors has yet had an abnormal child because of the rays" (17).

24. More from Gerstell 1950: "It is not *true* that the atomic bomb will make the whole world unfit live in" (18; italics in original). "It is not *true* that the explosion of atomic bombs sooner or later means the end of all life on earth" (ibid.; italics in original). "Scientists say it would take almost a *million atomic bombs all exploded a very short time* to 'doom' the earth. . . . So don't worry about *that*" (115; italics in original).

25. Boyer 1994: 22, 12.

26. *Collier's,* special issue, "Preview of the War We Do Not Want," October 27, 1951.

27. Rockefeller 1960: 413.

28. When the USSR blockaded Berlin in 1948, the United States not only airlifted supplies to the city, it also sent B-29s to Britain. Oakes (1994) writes, "every policy maker in the East and the West knew that the Superfortress was the aircraft that had dropped the atomic bombs on Japan" (12).

29. Speech delivered January 12, 1954; cited by Lifton and Falk 1982: 158.

30. U.S. Congress, Joint Committee on Atomic Energy 1959: 953.

31. Rockefeller 1960: 412–13.

32. Ibid.: 413.

33. Gerstell 1950: 1, 5, 9.

34. Ibid.: 10, 31, 32.

35. Ibid.: 14.

36. Ibid.: 27, 50.

37. Ibid.: 51.

38. Ibid.: 67.

39. In *Survival under Atomic Attack,* published by the U.S. Federal Civil Defense Administration in 1950, one finds another excellent example of this kind of effort to convince the public that the atomic bomb is pretty much just a big bomb. A few simple measures will protect you from blast and heat. Radioactivity is not much of a problem either. If you are exposed, simply dispose of your clothes and wash with soap and water. If you are contaminated, you could get sick to your stomach, vomit and "feel below par . . . your hair might fall out." The good news is that "you would still stand better than an even chance of making a complete recovery, including having your hair grow in again." According to Oakes (1994: 52) and Weart (1987: 13), twenty million copies of *Survival* were printed and distributed.

40. Paul Nitze, "Reexamination of the United States Program for National Security," NSC[National Security Council]-141, quoted by Kaplan 1983: 138.

41. The break between the two approaches was not quite as clean as I am depicting it here. As late as 1955, some of the FCDA's publications

were still claiming that getting through an atomic attack was pretty much like surviving a tornado.

42. Oakes 1994: 62.

43. Ibid.: 62–63. A look at some comments by important policy opinion leaders confirms Oakes's analysis. Nelson Rockefeller: "If another Berlin comes along, . . . [w]ill we be able and ready to stand, unless we know we would be able to survive? We would not, under present circumstances. . . . We would, if we had shelters. . . . shelters can play [a role] in strengthening the hand of our government . . . [shelters] can affect a major psychological change on the American people, . . . we can hold our heads higher . . . without shivering every time somebody mentions the possibility of a nuclear attack" (1960: 412, 413). Paul Nitze, the Cold Warrior who would still be a key figure in foreign policy circles three decades later in the Reagan administration, made the same point in NSC-141: "[The] willingness of the United States . . . to initiate an atomic attack . . . will be significantly affected by the casualties and destruction which the Soviet system could inflict in retaliation. . . . the United States might find its freedom of action impaired. . . . [The prospect of heavy casualties] would tend to impose greater caution in our cold war policies . . . the freedom of the United States Government to take strong actions in the cold war . . . depend[s] in increasing measure upon firm public morale in the United States. . . . [Therefore, an] adequate civil defense is of utmost importance"; quoted in Kaplan 1983: 137–38.

44. "A Shelter in Time Saves Thine and Firms Up National Will," *Life* 51, 5 (August 4, 1961): 44.

45. Oakes 1994: 82.

46. U.S. Federal Civil Defense Administration, n.d. See J. Brown 1988, for a description of the whole of the civil defense campaign aimed at schoolchildren.

47. The same sequence, blown up, was displayed in the FCDA's traveling exhibit (Oakes 1994: 134); it also appeared in the March 30, 1953, issue of *Life*, 21–22.

48. Weart 1987: 13.

49. V. Peterson 1953: 100.

50. Weart 1987: 14–17; Oakes 1994: 84–104.

51. Eisenhower press conference, March 31, 1954, quoted in Williams and Cantelon 1984: 113.

52. *New York Times,* April 1, 1954.

53. Weart 1988: 183.

54. Officials had tried to persuade the public that fallout was hardly more than "an uncommonly troublesome form of household dirt" (Oakes 1994: 122). *Life* had this advice: "If you think you have been contaminated by fallout, remove all your clothing as soon as possible and wash

off your skin and especially your hair. . . . The best first aid for radiation sickness—whose symptoms are nausea, fatigue and fever—is to take hot tea or a solution of baking soda to combat the nausea and aspirin for the fever. You can recover from a mild case of radiation sickness just as you recover from a cold" (51, 11 [September 15, 1961]: 108). But this effort to dispel people's fear did not work. Ralph Lapp identified what it was about fallout that so terrified: "Blast can be readily felt as can heat . . . Radioactivity, on the other hand, cannot be felt and possesses all the terror of the unknown. It is something which evokes revulsion and helplessness" (Lapp 1954, reprinted in Williams and Cantelon 1984: 188).

55. U.S. Congress, Joint Committee on Atomic Energy 1959: 380–81.

56. Lapp 1954, reprinted in Williams and Cantelon 1984: 183; Winkler 1984: 18; Kerr 1983: 68–76.

57. Winkler 1984: 18.

58. Lapp 1954, reprinted in Williams and Cantelon 1984: 184.

59. U.S. Congress, Joint Committee on Atomic Energy 1959: Chart 4, inserted between pages 44 and 45; U.S. Department of Defense, Office of Civil Defense 1968: 82.

60. *Bulletin of the Atomic Scientists,* www.bullatomsci.org/issues/ nukenotes/nd97nukenote.htm.

61. The narrative in this section is based on Mitchell 1962; Fitzsimons 1968; Kerr 1983; and Winkler 1984.

62. Clarence Cannon, chair of the House Appropriations Committee, expressed the opinion "that *no* civil defense program, regardless of expense, could possibly do the job. . . . an attack upon the United States would be such a catastrophe that any protection provided by advance planning would be but 'a drop in the bucket'" (Kerr 1983: 55).

63. Peterson told a congressional committee that if Washington were hit by a hydrogen bomb, there would only be "three alternatives: die, dig, or get out" (ibid.: 63). Since Congress did not seem inclined to pay enough to dig, evacuation seemed the only alternative to dying.

64. Ibid.: 71.

65. Since Eisenhower was a fiscal conservative, military leaders assumed that money spent on civil defense would be money not spent on weapons. In 1960, Gen. Curtis Lemay, Commander of the Strategic Air Command, famously remarked at a congressional hearing, "I don't think I would put that much money into holes in the ground to crawl into[;] . . . I would rather spend more of it on the offensive weapons systems to deter the war in the first place" (ibid.: 109).

66. Mitchell 1962: 52, 24.

67. Ibid.: 18, 25. Criticism of the state of civil defense planning was bipartisan. Chet Holifield, the Democratic congressman who had paid the most sustained attention to such matters during the preceding ten years,

said, "civil defense . . . is in a deplorable state" (Kerr 1983: 116). James Tobin, a member of Eisenhower's Council of Economic Advisors, called civil defense in the United States "a joke" (Fitzsimons 1968: 39).

68. The fallout shelter panic gets its fullest treatment in Waskow and Newman 1962. See also Kerr 1983; Kaplan 1983; Watson 1984; Winkler 1984; and Weart 1987. The episode is also discussed in books about Jack Kennedy: Schlesinger 1965; Sorensen 1965; Reeves 1993.

69. Schlesinger 1965: 747.

70. Kerr 1983: 138–42; Weart 1987: 17–20. Leon Festinger's theory of cognitive dissonance may be helpful here. Festinger's notion is that people cannot tolerate "dissonance" among simultaneously held beliefs, feelings, or thoughts. Faced with dissonance, people are motivated to bring these potentially clashing psychic elements into harmony through a variety of psychological operations. Helplessness in the face of immense threat can, for example, be dealt with through denial, especially if the threat, no matter how dire, is a low probability one.

71. U.S. Congress Joint Committee on Atomic Energy 1959: 651, 251, 252, 255.

72. Ibid.: 437, 298.

73. Kahn 1961: 77.

74. Mitchell 1962: 56, 57.

75. U.S. Congress, Joint Committee on Atomic Energy 1959: 787.

76. Mitchell 1962: 56.

77. U.S. Congress, Joint Committee on Atomic Energy 1959: 786.

78. Ibid.: 785.

79. Ibid.: 788–89.

80. Ibid.: 790.

81. Ibid.: 789.

82. Ibid.: 834, 835.

83. Ibid.: 835.

84. Ibid.: 836, 838.

85. Fromm and Maccoby 1962: 15.

86. Mitchell 1962: 65.

87. Ibid.: 72.

88. Inside the White House, Eisenhower and his advisors talked of the probable need for military dictatorship to deal with the postwar situation. Oakes 1994: 155–56.

89. Kahn 1962b.

90. U.S. Congress, Joint Committee on Atomic Energy 1959: 894.

91. Wigner 1968: 23.

92. Kahn 1962a: 6; Kahn 1961: 77.

93. U.S. Congress, Joint Committee on Atomic Energy 1959: 791.

94. Waskow and Newman 1962: 35–36.

95. *Life,* 51, 11 (September 15, 1961).

96. Swayze 1980: 20.

97. Ibid.

98. DeArmond 1996.

99. Haitch 1983.

100. Waskow and Newman 1962: 13, 85.

101. Rozhon 1999.

102. Schlesinger 1965: 748; Reeves 1993: 271–72.

103. U.S. Congress, Joint Committee on Atomic Energy 1959: 689.

104. Martin and Latham 1963: 215–47.

105. Fromm and Maccoby 1962: 16.

106. Ibid.: 17. The psychological problems of shelter living are also discussed in Martin and Latham 1963: 271; and A. M. Katz 1982: 200–207.

107. Dentler and Cutright 1964: 412.

108. U.S. Federal Civil Defense Administration 1950: 25.

109. Oakes 1994: 163.

110. Ibid.: 164, 163.

111. Civil Defense Letter Committee 1961: 29.

112. Piel 1962: 2, 7.

113. Kahn 1961: 92, 91.

114. Fromm and Maccoby 1962: 20.

115. Ibid.: 16, 17, 18.

116. See descriptions of victims at Hiroshima in Hersey, cited by Dentler and Cutright 1964: 420–21. The potential psychological impacts of atomic war were explored with great sensitivity, later, by the psychiatrist Robert Jay Lifton. See Lifton, Markusen, and Austin 1984.

117. Oakes 1994: 150, 151.

118. Singer 1961: 311.

119. Fromm and Maccoby 1962: 18.

120. U.S. Congress, Joint Committee on Atomic Energy 1959: 838.

121. Commoner 1966: 102. By the 1980s, catastrophic ecological impacts were at the forefront of the case against nuclear war. See J. Peterson 1983; P. Ehrlich 1983, 1984; Howes and Ehrlich 1987. Carl Sagan and others popularized the term *nuclear winter.* Levi 1987; A. Ehrlich 1988. In his best-selling book *The Fate of the Earth* (1988), Jonathan Schell predicted that nuclear war would radically simplify nature, turning the earth into a "world of grasses and cockroaches." The ecological critique was both the most inclusive version of the interdependence argument and the most powerful argument that fallout shelters, even if they could save some lives in the short run, were truly a textbook model for why inverted quarantine does not—cannot possibly—work in the long run.

122. The phrase occurs both in the Civil Defense Letter Committee's "Open Letter" (1961): 29; and in Fromm and Maccoby 1962: 22.

123. Piel 1962: 8; Singer 1961: 314.
124. Singer 1961: 314.
125. Civil Defense Letter Committee 1961: 29; Fromm and Maccoby 1962: 22.
126. Civil Defense Letter Committee 1961: 29; Federation of Atomic Scientists 1962: 25, 27; Piel 1962: 8.
127. Singer 1961: 313.
128. Ibid.: 312; Piel 1962: 8.
129. Inglis 1962: 18.
130. Panofsky 1966: 18.
131. Ibid.
132. Chamberlain 1966: 28.

2. Suburbanization as Inverted Quarantine

1. All census figures from Hobbs and Stoops 2002: 32, 33. The actual percentage of the population that can be said to be suburban is likely quite a bit higher than the official statistic of 50 percent. Many communities that were originally built as suburbs were subsequently annexed by their central city (Kasarda and Redfearn 1996: 406; Jackson 1985: 138–56, 276–78). Residents of such places—suburblike in all respects but formally within city boundaries—are counted as urban dwellers in subsequent censuses. Also, many of the newer cities of the South and the West are (and have been from the start) essentially vast, sprawling aggregates of residential tracts. Although officially designated cities, these places are in every real sense suburbs, and their residents, though officially counted as urban dwellers, are really suburbanites. See Frey 2003: 162; and Fishman 1987.
2. Jackson 1985: 4.
3. Ibid.: 68.
4. Ibid.: 72.
5. Ibid.: 64.
6. I should clarify how I mean to use the term *suburb*. If one defines suburb as any community that lies outside an urban area and is functionally dependent on it (rather than being an autonomous and self-contained community), there were suburbs in the United States as far back as the early nineteenth century. They were industrial suburbs, working-class suburbs, sometimes just slums, not the kinds of communities we now associate with the term *suburb*, which evokes images of residential subdivisions with nice homes and lawns (Harris 1999). I am not interested in all suburbs, in any and all communities that ever existed on the urban fringe. I am interested in the development of the kind of communities that are like the popular image of the suburb, the "bedroom" suburbs or the "commuter" suburbs, embodiments of the Suburban Ideal.

7. Jackson 1985: 78.

8. Ibid.: 205.

9. Muller 1996: 401; Jackson 1985: 113, 183.

10. Ruben 2001: 440.

11. Brennan and Hill 1999; Glaeser, Kahn, and Chu 2001.

12. Garreau 1991; Gottdiener and Kephart 1991; Lang 2002.

13. Kling, Olin, and Poster 1991; Fishman 1987: 182–207.

14. Kling, Olin, and Poster 1991: 10.

15. Berube 2003a: 36, 41.

16. Lang 2002: 4. For recent (Census 2000) data on increasing diversity in the suburbs, see the Brookings Institution's Metropolitan Policy Program, Living Cities Databook Series.

17. Lucy and Phillips 2001, 2003: 132; Lang 2002; Firestone 1999; Holmes 1997. Lang (2002: 4) calls these fast-growing counties at the fringes of metro areas, "New Metropolitan Counties." (For more discussion of exurban growth trends, see Berube 2003a; Frey 2001, 2002, 2003.)

18. Lang 2002: 4. See also Lang and Simmons 2003.

19. The first gated communities were upscale retirement communities and "lifestyle" communities that combined upscale housing with "country club" facilities, that is, outdoor activities such as golf and tennis, in places like Florida, Arizona, and Southern California.

20. Blakely and Snyder 1997: 6; El Nasser 2002.

21. Blakely and Snyder 1997: 7. See also McGoey n.d.

22. Blakely and Snyder (1997) estimate 19,000 gated communities, 3 million households, and 8.5 million individuals living in these households by 1997. Analyzing the Census Bureau's 2001 American Housing Survey, Sanchez and Lang (cited by Low 2003: 15; and by El Nasser 2002) come up with very similar numbers: 4 million households in neighborhoods that had both walls and some form of controlled entry. That figure translates to about 9 million Americans living in walled and gated communities.

23. Tucker 1998.

24. Owens 1997: n. 1.

25. Cited by Lang and Danielsen 1997: 868.

26. Dillon 1994; Owens 1997; Lang and Danielsen 1997; Blakely and Snyder 1997; Tucker 1998; El Nasser 2002; Low 2003.

27. Jackson 1985 and Fishman 1987 were especially valuable. I found Muller's short geographic history of suburbanization (1996) also quite helpful.

28. In addition to Jackson 1985, see Hadden and Barton 1973.

29. Jackson 1985: 68.

30. Chevalier 1973, for the "dangerous classes" of Paris; Schupf 1971 and Symons 1849, for London; Brace 1967 [1880], for New York City.

31. Muller 1996: 395.

32. Jackson 1985: 46.

33. Ibid.: 48.

34. Ibid.: 59. Jackson quotes Andrew Jackson Dowling, an advocate of landscape gardening who became one of the central figures in the articulation of the connection between domesticity and the suburban ideal: to Dowling the individual family home has "power" and "virtue." Why? Because when "sensible men" come to their senses and "gladly escape . . . from the turmoil of cities," they will find waiting for them the "house in the countryside . . . the little world of the family home . . . [which is the place of] truthfulness, beauty and order" (ibid.: 64, 66). Other "pushes" and "pulls"—upward mobility and class emulation as "pull"; conversely, fear of downward mobility, the fear of status erosion because one lives in proximity to recent immigrants or to people of a lower class or of another race, as a "push"—reinforced and amplified the attraction of the suburb.

35. It is apt, here, to recall Thorstein Veblen's (1899) sardonic, insightful discussions of conspicuous consumption and of invidious comparison.

36. This is an early example of the "growth machine" dynamics described by Logan and Molotch 1987.

37. Ewen 1976.

38. Muller 1996: 403.

39. Ibid.

40. Jackson 1985: 157ff.; Muller 1996: 401, 402.

41. Jackson 1985: 196–215; Massey and Denton 1993: 51–54.

42. Jackson 1985: 207. Jackson (1985: 203–18) provides a detailed discussion of how FHA policy tilted resources away from city and toward suburb. Massey and Denton (1993: 51–54) focus specifically on the de facto racism of FHA policies.

43. Muller 1996: 402.

44. Jackson 1985: 206. The federal government's tilt toward the suburbs was structurally reinforced by how the geographic shift in population affected the distribution of congressional seats. As population shifted to the suburbs, the number of members of the House who represented suburban interests grew, and the number who represented city constituencies shrank. In 1973, 78 members of the House represented urban constituencies, and 88 represented suburbs; by 1993, the number of urban congressional districts had fallen to 67, while the number of suburban districts almost doubled, to 160. Gainsborough 2001: 168.

45. Muller 1996: 399.

46. Glenn 1973.

47. U.S. Census Bureau 2000.

48. Berube and Frey 2002: 4. That is just the average. In some places the difference is much larger. In Washington, DC, the city poverty rate is

20.2 percent; in the suburbs, 5.6 percent. In Philadelphia, city, 22.9 percent; suburbs, 6.2 percent. In Baltimore, city, 22.9 percent; suburbs, 5.4 percent. In Cleveland, city, 26.3 percent; suburbs, 6.7 percent. In Newark, city, 28.4 percent; suburbs, 6.8 percent. Ibid.

49. For a comprehensive and detailed discussion of barriers to black suburbanization, see Massey and Denton 1993, especially chapter 2, "The Construction of the Ghetto."

50. Goldsmith and Blakely 1992: 121.

51. Massey and Denton 1993: 61.

52. Goldsmith and Blakely 1992: 97.

53. Wilson 1987: 3.

54. Massey and Denton 1993: 2.

55. D. Massey 2001: 322.

56. Wilson 1987: 4.

57. Goldsmith and Blakely 1992; Massey and Denton 1993; Anderson and Massey 2001.

58. For discussions of joblessness, see Wilson 1987; Massey and Denton 1993; Madden 2001. For family and decreasing prospects for marriage, see Furstenberg 2001; Massey and Denton 1993 (e.g., "abandonment of marriage," 158); and Wilson 1987 (see "family dissolution," teen pregnancy, out-of-wedlock births, female-headed households [26]). For violent crime, see D. Massey 2001; and Sampson and Wilson 1995.

59. Massey and Denton 1993: 58.

60. Quoted by Massey and Denton 1993: 4.

61. Goldsmith and Blakely 1992: 97.

62. Ruben 2001: 444.

63. Simmons and Lang 2003.

64. Massey and Denton 1989; Charles 2001. Denton (1994: 49) writes that the term "describes the extreme, multidimensional, cumulative residential segregation experienced by African Americans in some large metropolitan areas, mainly in the Northeast and Midwest." Douglas Massey (2001: 319) says the term still applies in 2000.

65. Frey 1979: 427.

66. Melosi 1980, 2000.

67. Realistically, no amount of reform could have transformed cities to the point that they offered *all* the amenities promised by the suburbs. Most city dwellers could never have the lawn, the square footage, or the distance from the neighbors promised by the boosters of suburban living. For that you need land that is cheap and plentiful. But if one's definition of the good life tilts toward other values—diversity, a rich and stimulating environment, the ideals of urban life championed by, say, Jane Jacobs—rather than tilting toward the suburban ideal, the example of some Western European cities shows that with sufficient reform and investment cit-

ies *can* offer their residents many of those amenities, while minimizing the problems we in the United States associate with city living.

68. Lucy and Phillips 2003. See also Fasenfest, Booza, and Metzger 2004; Jargowsky 2003; Bier 2001; and Lang and Danielsen 1997.

69. Berube and Frey 2002: 4.

70. Ibid.

71. Frey 2002. See also Frey 2001; and Frey 2003: 159, 165.

72. Lang 2002. Firestone (1999) describes an exurban county in Georgia where 99.3 percent of the population is white.

73. Frey 2002. Frey's article is based on analysis of the 2000 census. Updating, the Census Bureau reports in 2006 that exurban communities continue to grow, with rates that exceed other communities in the nation (Lyman 2006).

74. Holmes 1997.

75. The fear/security theme comes through loud and clear in the best ethnographies of the gated community: Low 2003; Blakely and Snyder 1997.

76. In addition to Low 2003 and Blakely and Snyder 1997, see Lang and Danielsen 1997; Dillon 1994; Owens 1997; Egan 1995; and Diesenhouse 1996.

77. Low 2003: 57, 70.

78. Blakely and Snyder 1997: 77.

79. The largest numbers of gated communities are found in states with "Melting Pot" (Frey 2003) metros such as Los Angeles, Oakland, Anaheim, San Diego, Riverside, Dallas, Houston, and Miami. The 2000 census confirms that these metros now have large minority populations in their *suburbs*, not just in their inner cities. In the metros named here, racial and ethnic "minorities" make up from 33 percent to 78 percent of their suburban populations (Brookings Institution, Metropolitan Policy Program, Living Cities Databook Series).

80. Brookings Institution, Metropolitan Policy Program, Living Cities Databook Series; Berube and Frey 2002.

81. The two articles are Davis 1992 and Flusty 1997.

82. Eyewitnessed by the author, walking around Encino, an upscale neighborhood in "the Valley" in Los Angeles.

83. Davis 1992: 174. The National Institute of Justice reported there were 1.6 million private security guards nationwide in 1994 (Owens 1997). Owens (1997) predicted that that number would be up to 1.9 million by the year 2000, and he cites the Department of Justice estimate that by 2000 the private security sector would be a $104 billion industry.

84. Davis 1992: 154, 155.

85. Flusty 1997: 53.

86. Davis 1992: 159.

87. Flusty 1997: 53.
88. Boddy 1992: 125.
89. Davis 1992: 160, 163. Flusty (1997: 54), a bit more sardonically, describes the public art in these places as "high-art plaza-turds signed by some of the best plop-artists."
90. Boddy 1992: 135–36.
91. Davis 1992: 159.
92. Ibid.
93. Tavernise 2004.
94. Boddy 1992: 138.
95. Davis 1992: 161–63; Flusty 1997: 49. For anyone interested in what one might call the semiotics of urban architectural paranoia, Flusty's essay is a must read.
96. Davis 1992: 155.
97. DeVries and Dorfman 1980: 22; Skonsen 1998.
98. These quotes are from the cover of the October 1998 issue of *American Survival Guide*.
99. Text on the cover of *New West*, February 25, 1980, advertising DeVries and Dorfman's (1980) article on survivalists.
100. I wish to note that I am not, here, interested in the most familiar and common critique of suburban life, that suburban living is too comfortable, too safe, too predictable, that it lacks those qualities that made city living so good for the soul, the city's messiness, disorder, vitality, diversity. See, for example, Mumford 1961, Jacobs 1993 [1961], and Sennett 1970. Personally, I think that critique is elitist, snobbish, too one-sided, too much of a stereotype. Plenty of people have grown up in the suburbs and gone on to live productive, full, interesting, creative lives.

I am more sympathetic to the more recent, *ecological* critique of suburbanization. Suburban sprawl, with its subdivisions, roads, parking lots, malls, consumes and fragments habitat. The suburban built environment is extravagantly wasteful of energy and natural resources. A freestanding suburban home requires lavish amounts of materials to build and furnish. It gobbles up energy for lighting, heating, cooking, washing, air conditioning. Lawn maintenance requires a lot of water (and, of course, the application of weed killer and fertilizer). The social geography of the suburb radically separates living from working from shopping, thus perpetuating our dependence on the automobile, which, as we are now so aware, is at the epicenter of a host of environmental ills, ranging from the extraction of crude oil to engine emissions that degrade air quality and contribute to climate change. For summaries of ecological critique, see Sawyer 2002 and Cieslewicz 2002. Important and valid as I think this critique is, it is tangential to my main line of inquiry here.

101. Noted by Muller 1996: 403.

102. Goldsmith and Blakely 1992: 121.

103. Gans 1972: 88.

104. Some recent developments have been taken as evidence that this pattern of separate and unequal social worlds may be breaking down somewhat. Demographic data suggest that suburbs have become more racially or ethnically diverse in recent years. Cities are growing again. New development is revitalizing some inner cities. Demographers even detect some movement of whites back downtown. Though real, I do not think these developments have significantly altered the essential features of the social geography I have been describing.

The suburbs have grown more diverse. Increased diversity does not mean greater integration, however. Logan (2003: 253) writes that "newly suburban group members [have] tended to move into the same array of neighborhoods in which co-ethnics were already living in 1990." As a result, "minority segregation and isolation have increased in the suburbs during the 1990s as suburbs have become more diverse" (246). At the same time, whites continued to retreat into gated communities or flee to new exurban subdivisions, especially exurbs in the "New Sunbelt" (Frey 2002), where "the fastest growing counties in the U.S. . . . . counties on the peripheries of New Sunbelt metro areas . . . are largely white" (ibid.).

Turning to the cities, is it true that the trend of urban decline has been reversed? Cities are growing again, rebounding from population losses during the 1970s, but looking only at total numbers misses important trends. Dramatic increases in the sizes of urban Asian and Hispanic communities account for most of the growth (Berube 2003b; Glaeser and Shapiro 2003; Simmons and Lang 2003; Frey 2003). These increases mask continued white flight from the cities. Berube (2003b: 143) reports that the biggest one hundred cities lost a combined 2.3 million whites during the 1990s, an outmigration that is far greater than the reverse trend that brings some young white urban professionals and some white suburban "empty nesters" back into "revitalized" inner cities. Segregation in the cities may have decreased a bit, by some measures, but the bottom line is that "truly staggering levels" of segregation still persist in most cities in the United States (Glaeser and Vigdor 2003: 226).

Furthermore, population growth is a positive indicator of societal health only if it is accompanied by economic development that generates good jobs. Some cities have experienced job growth, but jobs in these cities' suburbs continue to grow at faster rates. About a quarter of big cities actually lost jobs in the mid-1990s, a time of vigorous national economic expansion (Brennan and Hill 1999). Some cities have revitalized their downtowns with new convention centers/hotel complexes, high-rise buildings for corporate offices, sports arenas, new or updated museums, and, of course, so-called festival marketplaces that turn abandoned, deteriorating industrial districts into tourist destinations. This kind of redevelopment brings tourists to the city and, the demographers tell us, even a "trickle" of white residents returning to live downtown (Sohmer and Lang 2003), but they tend to create mostly low-paying service sector jobs, such as restaurant work or retail sales. In effect, this kind of downtown "revitalization" tends to create "pockets of gentrification" (Ruben 2001: 435) that coexist with and do nothing to improve conditions in deeply distressed inner city neighborhoods. (See also Katz and Bradley 1999.)

105. Quoted by Putnam 2000: 210.

106. Baumgartner 1988: 10.

107. Ibid.: 11, 107.

108. Duany, Plater-Zyberk, and Speck 2000.

109. Oliver 2001.

110. Putnam 2000: 210.

111. Kling, Olin, and Poster 1991: 8, 7.

112. Low 2003: 64.

113. Blakely and Snyder, cited by Low 2003: 23.

114. Tucker 1998.

115. Baumgartner 1998: 10.

116. Kling, Olin, and Poster 1991; Brooks 2002.

117. Brooks 2002, 2004; Firestone 1999; Holmes 1997; Lang 2002.

118. Brooks 2002.

119. In addition to the sources cited above in my discussion of affluent city dwellers forging protected spaces for themselves, see Charles 2001: 284.

120. P. L. Brown 1997; Rose 2001. A movie about the safe room, *Panic Room*, was released in 2002.

121. Oliver 2001; Gainsborough 2001; G. S. Thomas 1998.

122. Gainsborough 2001: 16.

123. Such generalizations, obviously, do not apply to *every* person who lives in suburbs. There, as everywhere, one always finds considerable individual variation with regard to any human trait. Nonetheless, it is appropriate to talk about a central tendency, or a quality possessed by a majority, or a kind of ideological Zeitgeist that preconsciously colors the perceptions, opinions, reactions, and values of every person who lives in a particular milieu. It is in that sense that it is valid to talk of a suburban social and political point of view, not universal but present clearly enough to leave its traces in opinion polls, voting behavior, or the policies of local officials, who must represent the views and opinions of their constituents if they wish to stay in office.

124. Thomas Byrne Edsall with Mary D. Edsall, *Chain Reaction: The Impact of Race, Rights, and Taxes on American Politics* (New York: Norton, 1991), quoted by Gainsborough 2001: 20.

125. G. S. Thomas 1998: 144; A. Cohen 2004.

126. Frey 2002.

127. Just to cite a few of the many articles and books that make this point: for the suburb, Baumgartner 1988; for postsuburban communities, Kling, Olin, and Poster 1991; for gated communities, Blakely and Snyder 1997; and Low 2003; for upscale, fort-ified city places and the urban zones that surround them, Davis 1992; Boddy 1992; and Flusty 1997.

128. Muller 1996: 403.

129. Baumgartner 1998: 3, 134, 131, 132, 134.

130. G. S. Thomas 1998: 152, 151.

131. Oliver 2001: 5.

132. Residents of a separately incorporated jurisdiction gain in two ways: taxes maintain high quality of local services; at the same time, taxes are lower than they would be if those taxes flowed to a central city government and were used to address the more severe, therefore more costly, problems of the central city (pointed out by Massey 2001: 337). Like those Russian nesting doll toys that open up to reveal ever-smaller dolls inside, the logic of defensive localism can be deployed over smaller and smaller geographic scales, as one's definition of what constitutes "our community" shrinks. First, suburban areas incorporate. Subsequently smaller entities, homeowners' associations (HOAs), have argued that association dues maintain parks, streets, facilities that are within association boundaries but still open to general public use, so it is tantamount to "paying twice" if local government then levies property taxes that are used to maintain city, county parks, streets, and so on. They want to deduct HOA dues from their property taxes and/or state and federal taxes (Stark 1998; Kennedy 1995). In this way, HOAs are trying to redraw the boundary between public and private, further eroding the social ethic that says citizens should contribute to the common good, promoting, instead, the property-based anti-ethic that privileged communities should have the right to withdraw into their privatized, pampered spaces and everyone else should stop asking them for anything and go fend for themselves. What makes all of this more than just of passing interest is the continuing proliferation of HOAs: 10,000 in 1970; 55, 000 in 1980; 130,000 in 1990; by one estimate 230,000 by 2003 (Kennedy 1995; Low 2003: 177).

133. Oliver (2001) finds low rates of political participation in American suburbs. Low levels of participation do not necessarily mean that residents are apolitical, however. It can mean that a community has already successfully erected *structural* barriers, such as incorporation, that ensure that the political system will routinely, without further need for involvement, reliably deliver policies they like. If there is not much at stake, if the community has plenty of resources and few internal problems, and

is shielded from external problems and demands, why bother to take the time to talk with neighbors, attend city council meetings? One can still show up if something important is at stake. Oliver's findings are not, then, inconsistent with Goodall's (1972) finding that suburbanites do participate in local government when something that is important to them, something about local schools or something that could affect property values, is on the agenda.

134. Jackson 1985: 276–77.

135. Gans 1972: 86, 87.

136. Phillips 1972: 175.

137. G. S. Thomas 1998: 162.

138. Ibid.: 152, 151. For more evidence supporting this point, see ibid.: 61–67, 150–52, 160, 162; Gainsborough 2001: 16, 74–76, 93; and Gayk 1991: 285.

139. G. S. Thomas 1998: 140.

140. Phillips 1972: 179.

141. Gainsborough 2001: 168. G. S. Thomas (1998: 72) estimates that as of 1991, 40 percent of the House of Representatives, half the Senate, and 60 percent of the electoral college represented areas dominated by the suburban vote.

142. Gainsborough 2001: 16.

143. Ibid.: 4, 71–73.

144. Ibid.: 23, 111.

145. Connelly 2004.

146. Lang 2004.

147. G. S. Thomas 1998: 95. The resident of the gated community is not completely safe, however. Walls can be scaled, gates neglectfully operated. They do not keep out the person intent on getting in and doing mischief. Crime rates in gated communities are lower than urban rates but are similar to rates in nongated neighborhoods in the same general area. Blakely and Snyder 1997; Low 2003; McGoey n.d.

148. Inverted quarantine, voluntary withdrawal into protected space, can trigger a new kind of positive feedback loop that we might label the "hypersensitivity effect"; increasingly intense inverted quarantine activity provides higher and higher levels of partial protection but also triggers increasingly intense feelings of vulnerability, of being at risk. Something like this is seen, according to researchers, in residents of gated communities. Those residents are by and large less likely to fall victim to crime than other Americans, but paradoxically, perhaps ironically, they continue to be obsessed with crime. They seem, if anything, even *more* afraid than they were before they retreated behind the walls and the gates. Blakely and Snyder 1997; Low 2003: 21, 123.

149. I do think the term does describe conditions in certain parts of the social world, if not yet the whole of it. Davis (1992: 176–77) uses

carceral to describe social conditions in Los Angeles today, and, given his description of the city, I cannot argue with his use of the term.

150. The phrase is inspired by the title of Nan Ellin's *Architecture of Fear* (1997), and by the title of one of Mike Davis's books about Los Angeles, *Ecology of Fear.*

Introduction to Part II

1. "Trade Secrets."

2. U.S. Department of Health and Human Services 2001.

3. Richard Levinson, associate executive director of the American Public Health Association, commented that "there are 80,000 chemicals in commerce today, but only 27 chemicals in today's report"; quoted in Lazaroff 2001b.

4. Quoted in ibid.

5. U.S. Department of Health and Human Services 2003.

6. Pegg 2003; Schafer et al. 2004. Both also report that the U.S. Department of Health and Human Services CDC 2003 data show that Hispanics' bodies had significantly higher concentrations of certain pesticides.

7. U.S. Department of Health and Human Services 2005.

8. Physicians for Social Responsibility 2005.

9. Environmental Working Group 2003; World Wildlife Fund 2004; Commonweal Biomonitoring Resource Center 2005.

10. Environmental Working Group 2005. In 2006, the Associated Press reported a new study that found PFOA in the cord blood of 298 of 300 babies tested. Associated Press 2006.

11. Roan 2003; Lunder and Sharp 2003; Environmental News Service 2005b.

12. Examples of this process—how improvements in research methodology typically lead regulators to keep lowering what is considered a safe level of exposure—can be found in chapter 3.

13. Steinemann 2005; Waldman 2005.

14. Several studies showing this unexpected effect—health effects at low doses when there are none at higher doses—are cited by Steinemann 2005.

15. The classic case of synergist effect is the powerful combined impact of asbestos exposure and cigarette smoking on lungs. Some other examples of synergistic effects have come to light, even though few such interactions have ever been studied. See the discussion in chapter 3.

3. Drinking

1. *Beverage Industry Annual Manual* 1986: 10; *Beverage Industry Annual Manual* 1987: 45–46.

2. *Beverage Industry Annual Manual* 1989/90: 47.

3. Ibid.

4. *Beverage Industry Annual Manual* 1988: 47, quoting from Beverage Marketing Corp., "U.S. Bottled Water Market and Packaging Report, 1987." The quoted report mentions other, secondary reasons: "'concerns for personal health, fitness and weight control as well as alcohol moderation'"; ibid.

5. Water Quality Association, "Perceptions about Household Water," March 17, 1999, www.wqa.org/sitelogic.cfm?ID=88 (accessed November 20, 2001); Water Quality Association, "2001 National Consumer Water Quality Survey Fact Sheet," press release, April 23, 2001; Water Quality Association, "Eighty-Six Percent of Americans Have Concerns about Their Home Drinking Water, New WQA Survey Finds," press release, April 23, 2001.

6. Consumption passed five gallons per capita in 1985, 10 gallons per capita in 1993. By 2003, bottled water consumption had passed consumption of juices, coffee, beer, and milk and had become the nation's second most popular drink; International Bottled Water Association (IBWA), "U.S. Bottled Water Market Volume, Growth, Consumption 1976–1999," www.bottledwater.org/public/volumegrowthand consumption.htm (accessed April 16, 2006); IBWA, "Solid Gains Put Bottled Water in No. 2 Spot," bottledwater.org/public/downloads/2004/2003_BW_Stats_for_Web.doc (accessed April 16, 2006).

7. Official statistics gathered by the Centers for Disease Control and Prevention (CDC) are not useful for assessing the risk. Even in the case of disease caused by bacteria in water, the CDC says its statistics "probably underestimate the true incidence" of waterborne disease outbreaks (Centers for Disease Control and Prevention 2000: 12). Ford (1999) estimates that the actual incidence of waterborne illness is probably a whole order of magnitude greater than the official CDC numbers. The CDC also says that it has trouble counting cases of illness caused by chemicals in drinking water: "Waterborne chemical poisonings are probably underreported to CDC. . . . Exposure to chemicals via drinking water can a) cause illness that is difficult to attribute to chemical intoxication or b) cause nonspecific symptoms that are difficult to link to a specific chemical. . . . Physicians might have difficulties recognizing and diagnosing chemical poisonings" (Centers for Disease Control and Prevention 2000: 15). If the CDC cannot reliably track the incidence of frank chemical intoxication, i.e., cases where high-level exposure produces immediate, obvious, sometimes incapacitating symptoms, its figures surely do not tell us anything about the thing Americans worry about most, chronic exposure to low doses of toxic chemicals.

8. "Environment: The Cities: The Price of Optimism," *Time*, August 1, 1969: 41.

9. U.S. Environmental Protection Agency (EPA) 2000a: 58. Assess-

ment is based on the designated use for that water. Is the water a source for drinking water? Is it designated for recreational use only? Depending on the designated use, different standards are applied to determine if the water is "good" or "impaired"; ibid.: 53. The assessment is incomplete; only 23 percent of the nation's river-miles were assessed for this report; ibid.: 51. The EPA states that one cannot generalize from the 23 percent because that 23 percent is not a random sample. Furthermore, even for that 23 percent that the EPA declares "assessed," the quality of the data is very uneven. The assessment for half those miles was "based on qualitative information or monitoring information more than 5 years old"; ibid.: 52.

10. Ibid.: 84. Forty-two percent of all lake-acres were assessed for the EPA report. The EPA cautions that the condition of the assessed 42 percent may not accurately represent the condition of the remaining 58 percent; ibid.: 82.

11. U.S. EPA 2000b: inside front cover, unnumbered abstract page.

12. U.S. EPA 2000a: 158.

13. "Few states have the economic resources to assess ground water quality across an entire state. . . . large gaps in coverage exist. The data submitted by states . . . preclude a comprehensive representation of ground water quality in the nation at this time but, more importantly, may result in a skewed characterization of ground water quality that is more positive than actual conditions"; ibid.: 188–89.

14. Ibid.: 160. Research done by the U.S. Geological Survey finds, for example, that groundwater in many parts of the United States is contaminated with volatile organic chemicals, VOCs, a category of pollutants that includes MTBE, tetrachloroethene, trichloroethene, and trichloromethane (Squillace et al. 1999).

15. U.S. Geological Survey (USGS) 1999: 21. See also U.S. EPA 2000a: 70.

16. USGS 1999: 6, 8.

17. U.S. EPA 2000a: 158–59.

18. Ibid.: 167, 160. For a more recent USGS assessment of pesticides in American streams and groundwater, see Gilliom et al. 2006.

19. USGS 1999: 19.

20. U.S. EPA 2000a: 91.

21. For beef, I have seen estimates that range from four to as high as ten pounds of feed to produce one pound of meat.

22. U.S. GAO 1999b: 1, 3.

23. That estimate, for 1989, comes from the American Chemical Society (Szasz 1994: 11).

24. Atlas (1995) writes that "TRI data are a minuscule and severely skewed selection of all emissions . . . less that 1 percent of all wastes generated" (28).

25. U.S. EPA 2001a: E-2, E-6.

26. See Szasz 1986a: 6; Environmental Research Foundation 1987.

27. U.S. EPA 2000a: 160, 164. The most common industrial materials found in groundwater are metals, VOCs, semivolatile organic chemicals, and petroleum compounds; ibid.: 165–68.

28. Squillace et al. 1999: 4179, 4176 (emphasis added).

29. U.S. EPA 2000a: 88–89.

30. Ibid.: 195, 88.

31. U.S. EPA 2006a: B-1.

32. See Tarr 1996; and one or more of Melosi 1980, 1981, 2000.

33. U.S. EPA 2000a: 62, 88.

34. Ibid.: 165.

35. Squillace et al. 1999: 4176.

36. USGS 1999: 10.

37. Ibid.

38. See Szasz 1986a: 6; Environmental Research Foundation 1987.

39. U.S. EPA 2000a: 71. Other studies have shown that leachate from municipal landfills is as toxic as leachate from industrial hazardous waste landfills (Environmental Research Foundation 1988).

40. Spotts 2002.

41. U.S. EPA 2000a: 71.

42. Kolpin et al. 2002.

43. Ibid.: 1202.

44. Murphy et al. 2003. The study did find that these substances were present in concentrations that met SDWA standards. Water treatment further reduced the number of substances present, as well as the concentrations of the substances that remained, the researchers found, but the treatment did not remove the substances completely. Many VOCs were detected in the "finished," that is, water that had been treated and was officially ready to drink, samples.

45. Buckley 2004.

46. U.S. EPA 2000a: 202.

47. Barzilay, Weinberg, and Eley 1999: 26.

48. Ibid.: 25.

49. U.S. EPA 2000a: 210–11. The Safe Drinking Water Act exempts wells that serve fewer than twenty-five persons from having to comply with federal drinking water standards (U.S. EPA 1999a). One in ten Americans get their water from such small systems. The exemption affects rural communities disproportionately. Groundwater is the source of drinking water for 99 percent of the rural population, compared to 46 percent for the nation as a whole (U.S. EPA 2000a: 58). The concern here is that pesticides, herbicides, nutrients, and bacteria, as well as other contaminants, are common in groundwater in agricultural areas.

50. Bowers 1999.

51. Infrastructural problems are predicted to grow worse in the future, requiring massive spending to keep the system in relatively good repair—a point I will be returning to in chapter 7.

52. For a general description of how the act works, see U.S. EPA 1999a.

53. The USGS scientists who have looked at VOCs in groundwater wrote, for example, "Most of the detected VOC concentrations were less than current drinking water criteria; however, this should not be seen as a definitive appraisal of the health risks for several reasons. Drinking-water criteria are based on current toxicity information, and as new information becomes available, these criteria may be revised downward" (Squillace et al. 1999: 4186).

54. Maugh 2003.

55. U.S. GAO 1999c: 4.

56. Ibid.; U.S. GAO 1999a: 8.

57. Description of events surrounding the promulgation and withdrawal of the new arsenic standards are based on "Arsenic Water Standard Challenged by Wood Preservers," Environmental News Service 2001b; C. Fox 2001; Jehl 2001; Lazaroff 2001a, 2001c; R. Massey 2001.

58. R. Massey 2001.

59. USGS 1999: 9.

60. Squillace et al. 1999: 4186.

61. Kolpin et al. 2002.

62. For a good summary of health problems that can be attributed to polluted or contaminated water, see Barzilay, Weinberg, and Eley 1999, chapters 5, 6, and 8, pp. 60–89, 105–20.

63. U.S. GAO 1999c: 3. See also U.S. EPA 2000c.

64. U.S. GAO 1999c: 4.

65. U.S. EPA 2001b.

66. U.S. GAO 1999c: 11–12.

67. U.S. EPA 2005a, 2005b.

68. Even when "sound science" has shown beyond any doubt that a substance has quite serious adverse health impacts, officials may shy away from setting the standard as high as the science would suggest. Consider the example of chlorination by-products. The chlorine used in water processing combines with organic materials in the water to form a large variety of compounds known as organochlorides, among them the trihalomethanes (chloroform is the best known of the trihalomethanes). Although "most of these [disinfection by-product] compounds have not been identified to date, and those that have been identified have not been studied extensively," trihalomethanes have been linked to cancers of the bladder, liver, kidney, colon, and rectum, plus a number of other

reproductive and developmental disorders (Barzilay, Weinberg, and Eley 1999: 74). See Andelman, Meyers, and Wilder 1986; Barzilay, Weinberg, and Eley 1999: 74–77; Boorman et al. 1999; Weisel and Jo 1996; Weisel et al. 1999. The 1996 SDWA amendments called on the EPA to regulate disinfection by-products (U.S. GAO 1999a: 6). In response, the EPA issued regulations in 1998 that would, if fully implemented, "reduce exposure to disinfection by-products by 25 percent" (U.S. EPA 1999b: 3). That was an improvement, but it left the public well shy of being fully protected from this hazard.

69. USGS 1999: 6, 20.

70. Squillace et al. 1999: 4186.

71. Consider the findings of the Natural Resources Defense Council's examination of tap water in several California cities. All samples met SDWA standards, but Fresno's water contained nitrates, pesticides, and various industrial chemicals; Los Angeles's water contained disinfection by-products, arsenic, perchlorate, and nitrate; San Diego's contained disinfection by-products, perchlorate, ethylene bromide, and lead. Olson 2002.

72. Squillace et al. 1999: 4186.

73. Recall the biomonitoring studies reviewed in the introduction to Part II.

74. U.S. EPA 2000a: 167. In a recent contribution to the research effort, Hayes and colleagues (2006) assess the impact of chemical mixtures on amphibians.

75. Carpenter, Arcaro, and Spink 2002.

76. Barzilay, Weinberg, and Eley 1999: 26.

77. Nussbaum 2003.

78. "Introducing 16,000 Year Old Natural Geothermal Spring Water," flier distributed via the mail and with the product, 1995, Odwalla (Davenport, Calif.).

79. Text from "Island Water Floats on Luxury and Prestige," *Beverage Industry* 91, 9 (September 2000): 70–73; FIJI Water Company, "Bottled in Fiji; Shipped to You," 2005, http://www.fijiwater.com/bottled_in_fiji .html (accessed July 23, 2006).

80. From a Crystal Geyser magazine ad.

81. Big Sur Bottled Water, no date; emphasis (capitalization) in the original.

82. "Why Bottled Water?" www.ionics.com/AnnualReport/Bottled Water.htm. (accessed January 26, 2001). Text in brackets from Ionics 1995: 13, 15.

83. U.S. Congress 1991a: 97–98.

84. Barthes 1972.

85. Seen on NBC, May 4, 2002.

86. Swanson 1995. Pepsi and Coke later joined the effort to promote the hydration message. "Bottled Water Comes of Age," *Beverage Industry* 91, 9 (September 2000): 22–25.

87. Rockefeller University, "Americans Relate Water to Well-being, but Most Don't Get Their Fill," press release, May 30, 2000: 2.

88. Ibid.

89. National Environmental Education and Training Foundation 1999; Water Quality Association, "2001 National Consumer Water Quality Survey Fact Sheet," press release, April 23, 2001; Water Quality Association, "Eighty-Six Percent of Americans Have Concerns about Their Home Drinking Water, New WQA Survey Finds," press release, April 23, 2001.

90. National Environmental Education and Training Foundation 1999. The percentages exceed 100 percent because the survey allowed respondents to give more than one reason.

91. Yankelovich Partners 2000.

92. U.S. EPA 2003b: 5.

93. IBWA, "U.S. Bottled Water Market Volume, Growth, Consumption 1976–1999," www.bottledwater.org/public/volumegrowthand consumption.htm (accessed April 16, 2006).

94. *Beverage Industry Annual Manual* 1989/90: 49.

95. *Beverage Industry Annual Manual* 1989: 48.

96. *Beverage Industry Annual Manual* 1989/90: 49.

97. In 1999, four firms, together, had more than 50 percent of the market; "Bottled Water Industry Segmentation," www.bottledwaterweb.com, www.bottledwaterweb.com/indus.htm (accessed August 1, 2000). The number went from four to three in 2000 when Danone, the fourth biggest firm, bought McKesson, the third biggest.

98. *Beverage Industry Annual Manual* 1987, 1988, 1989; "Bottled Water Industry Segmentation," www.bottledwaterweb.com, www .bottledwaterweb.com/indus.htm (accessed August 1, 2000); "Bottle Water Soars," *Beverage Industry* 91, 9 (September 2000): Special Section, NP26-NP27.

99. www.perrierusa.com, click on "about us," then "our history," then "1992," (accessed July 8, 2006); www.bottledwaterweb.com/news/ nw_111798.html; www.bottledwaterweb.com/news/nw_011200.html; www.bottledwaterweb.com/indus.html.

100. www.bottledwaterweb.com/news/nw_053000.htm; www .bottledwaterweb.com/indus.html.

101. Beverage Marketing Corporation of New York, "News Release: Bottled Water Moves Up in the Rankings, Says Beverage Marketing Corporation," www.bottledwater.org/public/BottledWater2003.doc (accessed April 16, 2006).

102. See, for example, IBWA, "Demographics of the Bottled Water Consumer 1999," 2000, www.bottledwater.org/public/BWFactsHome_main.htm (accessed January 20, 2001).

103. Simmons Market Research Bureau 1992: P15-0002.

104. "Bottled water experienced the greatest success at higher-check establishments. . . . bottled water is consumed [in restaurants] most often by people in households with incomes of more than $40,000"; Gardner 1997. And no wonder: in New York, some restaurants charge $6, $10, even $12 for a bottle of Evian or San Pellegrino (Witchel 2000).

105. "Island Water Floats on Luxury and Prestige," *Beverage Industry* 91, 9 (September 2000): 70–73.

106. Simmons Market Research Bureau 1992: P15-0002.

107. von Wiesenberger 1999.

108. Water Quality Association, "Perceptions about Household Water," March 17, 1999, www.wqa.org/sitelogic.cfm?ID=88 (accessed November 20, 2001).

109. Water Quality Association, "2001 National Consumer Water Quality Survey Fact Sheet," press release, April 23, 2001; Water Quality Association, "Eighty-Six Percent of Americans Have Concerns about Their Home Drinking Water, New WQA Survey Finds," press release, April 23, 2001.

110. Water Quality Association, "Perceptions about Household Water," March 17, 1999, www.wqa.org/sitelogic.cfm?ID=88 (accessed November 20, 2001); Water Quality Association, "2001 National Consumer Water Quality Survey Fact Sheet," press release, April 23, 2001.

111. USA Today/CNN/Gallup Poll, June 5–7, 1998.

112. Water Quality Association, "Perceptions about Household Water," March 17, 1999, www.wqa.org/sitelogic.cfm?ID=88 (accessed November 20, 2001).

113. National Environmental Education and Training Foundation 1999.

114. U.S. EPA 2003b: 4.

4. Eating

1. Burros 1997: A1.

2. Organic Trade Association 2004; Organic Trade Association, "The Organic Industry," 2005d.

3. Organic Trade Association's 2004 Manufacturer Survey, cited by Klonsky and Greene 2005: 20.

4. Benbrook 2004: 4.

5. Yess, Gunderson, and Roy 1993; Benbrook 2004.

6. Yess, Gunderson, and Roy 1993; Foulke 1993.

7. Groth, Benbrook, and Lutz 1999; Benbrook 2004.

8. Foulke 1993.

9. Groth, Benbrook, and Lutz 1999: 2, 21–22. Note: All these different pesticides were not found on the same samples of apples or peaches; what this means is that when all the samples of, say, apples are considered together, one finds that thirty-seven different pesticides are applied to apple crops in various locations in the United States.

10. Benbrook 2004: 5.

11. Ibid.: 5, 12; Schafer and Kegley 2002: 813.

12. Benbrook 2004: 4.

13. Discussed by Groth, Benbrook, and Lutz 1999: 19–20.

14. Groth, Benbrook, and Lutz 1999: 3, 16.

15. U.S. GAO 2000: 7; U.S. EPA 2000a: 158.

16. Groth, Benbrook, and Lutz 1999: 31, 32.

17. Schafer and Kegley 2002: 813.

18. Chaney et al. 2003; Yang 2005; Peabody 2004; Durham 2004.

19. Littlefield 2004.

20. Ibid.

21. M. Fox 2006.

22. Ibid.

23. Becker 2004.

24. Pew Initiative on Food and Biotechnology 2004.

25. Thompson 2000.

26. See, for example, Organic Consumers' Association, http: //www .organicconsumers.org/gelink.html. Activists also point to studies that suggest a variety of potential *environmental* problems: GM plants cross-pollinating non-GM plants in nearby fields; GM genes "drifting," finding their way into other, wild species; pests becoming more resistant in response to GM crops that now produce their own pesticides. GM genes can contaminate organic crops (Environmental News Service 2003).

27. Becker 2004.

28. Pew Charitable Trust 2005.

29. Barboza 2000.

30. Whole Foods Market n.d.

31. Busboom and Penner 2002. These are the natural growth hormones; animals are also injected with synthetic growth hormones such as trenbolone acetate and zeranol.

32. Dupuis 2000.

33. Mellon, Benbrook, and Benbrook 2001: xi, xiii.

34. Busboom and Penner 2002.

35. European Commission 2000: 3, 2.

36. Ibid.

37. Commission of the European Communities 2000; National Cattlemen's Beef Association, "NCBA News: U.S. Cattlemen Call for EU to

Drop Beef Ban," press release, May 10, 1999, www.beefusa.org/NEWS USCATTLEMENCALLFOREUTODROPBEEFBAN4373.aspx (accessed May 9, 2006). Pressed by foreign exporters hurt by the ban, the commission reexamined its decision. Based on a thorough review of the most recent research results, the ban was reaffirmed (European Commission 2002).

38. That most legitimate of mainstream organizations, the American Medical Association (AMA), takes the official position that giving antibiotics to healthy farm animals is a grave concern because it will speed the evolution of resistant strains of bacteria. A link to the AMA resolution, passed in 2001, can be found at the Grace Factory Farm Project Web site: "The American Medical Association Approves Resolution to Eliminate Non-Therapeutic Use of Antibiotics in Agriculture," www .factoryfarm.org/docs/AMA.htm (accessed May 8, 2006). To see the range of the organizations mobilized against this practice, see the Keep Antibiotics Working.com's Web Site: Keep Antibiotics Working.com, www.keepantibioticsworking.com/new/index.cfm.

39. ConsumerReports.org, "You Are What They Eat."

40. Klonsky and Greene 2005: 2.

41. www.beef.org, "Myths & Facts about Beef Production: Pesticides in Food Production," http://www.beef.org/library/myths_facts/archive/myth_12.htm (accessed March 9, 2002).

42. Dewailly et al. 1989. For a more complete discussion, see chapter 6.

43. USDA 2000.

44. Foulke 1993. This advice is repeated at several sites by the EPA and seems to be the official policy line. See, for example, U.S. Environmental Protection Agency, "Pesticides and Food: Healthy, Sensible Food Practices," http://www.epa.gov/pesticides/food/tips.htm (accessed May 10, 2006).

45. Boutrif 2000.

46. ConsumerReports.org, "You Are What They Eat."

47. Larson 2001.

48. Larson 2001; Parrot 2005; Van Eenennaam 2005.

49. European Union 2003, 2006.

50. Diamand, Bebb, and Riley 1999; Carroll 2003.

51. Pew Charitable Trust 2005.

52. ConsumerReports.org, "You Are What They Eat."

53. Ibid.

54. U.S. FDA 2003.

55. ConsumerReports.org, "You Are What They Eat."

56. Alabama Cooperative Extension Service 1995; Boyles et al. 1988; Damron 2002; Faries, Sweeten, and Reagor 1991; Guyer 1980; Meyer 1990; van Heugten 1997.

57. The quote is from Boyles et al. n.d.; other contaminants listed here are mentioned by Guyer 1980; Damron 2002; and Faries, Sweeten, and Reagor 1991.

58. Guyer 1980.

59. Alabama Cooperative Extension Service 1995.

60. U.S. FDA, "EAFUS: A Food Additive Database," www.cfsan.fda .gov/~dms/eafus.html (accessed September 30, 2005). EAFUS is an acronym for Everything Added to Food in the United States.

61. "Do Food Additives Subtract from Health?" *Business Week,* May 6, 1996, www.businessweek.com/1996/19/b3474101.htm (accessed September 30, 2005).

62. U.S. FDA, Center for Food Safety and Applied Nutrition, "Food Ingredients and Packaging: Background Information for Consumers," http: //www.cfsan.fda.gov/~dms/opa-bckg.html.

63. Center for Science in the Public Interest, "CSPI's Guide to Food Additives," www.cspinet.org/reports/chemcuisine.htm.

64. Statistics gathered by the Organic Trade Association (OTA), cited in OTA 2004, 2005b, 2005d; and in Klonsky and Greene 2005: 20. Other sources give slightly different, somewhat higher estimates. There appears to be some imprecision in adding up aggregate sales. In spite of those differences, there is no question—complete consensus—about the overall trend.

65. OTA's "2004 Manufacturer Survey," cited by Klonsky and Greene 2005: 20.

66. OTA 2005b.

67. OTA's "2004 Manufacturer Survey," cited by Klonsky and Greene 2005: 20.

68. Pollan 2001; OTA's "2004 Manufacturer Survey," cited by Klonsky and Greene 2005: 20; Horovitz 2004; OTA, "OTA Survey Finds Organic Snack Food Sales Eating Away at Conventional Counterparts," news release, July 8, 2004, http://www.ota.com/news/press/147.html (accessed April 26, 2006).

69. OTA 2005c; Whole Foods Market n.d.

70. Pollan 2001; Chui 2000; Horovitz 2004; Schmeltzer 2005; OTA 2005c; Ness 2006.

71. Pollan 2001; also Chui 2000.

72. Kortbech-Olesen and Larsen 2001; Pollan 2001: 33; Shapin 2006.

73. Dupuis 2000: 285.

74. Pollan 2001: 32.

75. Results from this Hartman Group 2000 survey are reported at Agro Nanotechnology Corporation, "Organic Industry Background," www.natural-gro.com/research_org.htm (accessed April 24, 2006); and by Klonsky and Greene 2005: 7.

76. Whole Foods Market n.d.

77. Klonsky and Greene 2005: 7.

78. Pollan 2001: 35.

79. Ibid.

80. Agro Nanotechnology Corporation, "Organic Industry Background."

81. Klonsky and Greene 2005: 8–9. The authors do cite some exceptions. Apples can be grown organically at no extra cost (a fact reflected in the competitive price of organic apples), and other crops, too, under certain favorable conditions.

82. Klonsky and Greene (2005) cite two polls, one where 85 percent of the public agreed, the other where 91 percent agreed (8).

83. Cited in Agro Nanotechnology Corporation, "Organic Industry Background."

84. Kortbech-Olesen and Larsen 2001.

85. Blisard, Smallwood, and Lutz 1999.

86. Food Marketing Institute, "Spending on Organic Food Rises as American Consumers Seek to Balance Health and Nutrition Needs, according to FMI Study," press release, October 18, 2001, www.fmi.org/media/mediatext.cfm?id=362 (accessed October 31, 2001); Whole Foods Market n.d.

87. Food Marketing Institute, "Spending on Organic Food Rises."

88. Klonsky and Green (2005) cite a Roper Poll conducted in 2000: two-thirds of those who do not buy organic give as number one reason that organic is more expensive (7); the Whole Foods Market survey figure is 63 percent; Whole Foods Market n.d.

89. Whole Foods Market n.d. On the other hand, Whole Foods finds evidence that "income and education are not as important as they once may have been." In other words, as organic foods have gone mainstream, perhaps because a greater variety of organic foods is available and those items are increasingly sold in ordinary supermarkets, one does see increasing consumption of organic foods by people with more modest incomes.

5. Breathing

1. U.S. EPA 1997a.

2. U.S. EPA 1997b.

3. For the EPA's description of the health effects of smog and of particulates, see U.S. EPA 1997a, 1997b. Though the basic facts are known, new studies continue to clarify the risks. For lung cancer, see Nafstad et al. 2003. For reduced lung function in children, see Gauderman et al. 2004.

4. Brook et al. 2002; Pope et al. 2004; R. L. Johnson 2004.

5. U.S. EPA 1999c.

6. U.S. EPA 2000e: 1; U.S. EPA 2006b.

7. U.S. EPA 2000e: 3.

8. U.S. GAO 2006.

9. U.S. EPA 2000e: 66; U.S. EPA 2003a: 60; American Lung Association 2004, 2005, 2006.

10. Bell et al. 2004; Schneider 2004; Schneider and Hill 2005; "AMA and Two Dozen Other Health Groups Urge Strengthened PM Standards, 4-17-06," www.cleanairstandards.org/article/articleprint/472/-1/38/ (accessed May 25, 2006).

11. McCarthy 1999.

12. "AMA and Two Dozen Other Health Groups," www.cleanair standards.org/article/articleprint/472/-1/38/.

13. U.S. EPA and U.S. CPSC 1995.

14. Shore 1999: 1. Generally, the information in this section comes from Shore 1999 and U.S. EPA and U.S. CPSC 1995. I should also note that I am discussing only one facet of the indoor air problem, omitting discussion of other sources of indoor air pollution, such as radon, tobacco smoke, molds and mildew, combustion products from improperly maintained stoves, heaters, and fireplaces, application of pesticides indoors, and hazards due to now banned substances that may still be found in older homes, such as lead and asbestos (these are discussed in U.S. EPA and U.S. CPSC 1995).

15. U.S. EPA and U.S. CPSC 1995.

16. In addition to Shore 1999 and U.S. EPA and U.S. CPSC 1995, see Eichenseher 2004.

17. U.S. EPA and U.S. CPSC 1995.

18. Costner, Thorpe, and McPherson 2005: 11.

19. American Environmental Health Foundation, "About Our Products," www.aehf.com/store_html/aboutprd.html (accessed February 27, 2003).

20. Chemical Industry Archives, "Cancer in a Can: What the Chemical Industry Kept Secret about Vinyl Chloride in Hair Spray," www.chemicalindustryarchives.org/dirtysecrets/hairspray/1.asp (accessed June 12, 2006).

21. Union Carbide memo, 1973, cited in ibid.

22. The Environmental Working Group (2006) reports on the ingredients in more than fourteen thousand cosmetic and personal hygiene products. Of the almost seven thousand ingredients, only 11 percent have been tested. Thus, the fact that some of the ingredients are carcinogenic and some have reproductive effects (at least at higher concentrations) is swamped by the fact that almost nine of ten ingredients have not been tested.

23. Shore 1999: 5.

24. U.S. Department of Health and Human Services 2001, 2003, 2005.

25. Boorman et al. 1999; Weisel et al. 1999.

26. Tarr 1996.

27. Weisel and Jo 1996.

28. Boorman et al. 1999.

29. I am looking at some air filter ads in a Priorities catalog, and looking at an ad for The Sharper Image's Ionic Breeze GP Silent Air Purifier. U.S. EPA and U.S. CPSC 1995 confirms that most air purifiers do not (and do not claim to) remove hazardous chemicals from a home's air.

30. For a comprehensive, though now somewhat dated, review of this literature, see Szasz and Meuser 1997.

31. For example, see Steinman and Epstein 1995, and Berthold-Bond 1999.

32. Lists of such products may be found at the American Environmental Health Foundation's Web site, www.aehf.com, and, undoubtedly, many other Web sites.

33. Direct mail advertising from Kids Nature.

34. "Taste for Life," July 2003, inside front cover. The ads promise that if you use natural personal hygiene products, you will not inadvertently poison yourself while trying to be clean and attractive. Some products promise even more: they promise to actually *detoxify you,* help remove from your body hazardous materials acquired from other exposures (see Tannen 1994).

35. American Environmental Health Foundation's Web site, www.aehf.com.

36. Kahlenberg 2003; Preston 2003; Daley 2005.

37. La Ferla 2001.

6. Imaginary Refuge

1. California Legislature 1985: 2, 3.

2. Ibid.: 5.

3. Bonczek and Markussen 1987: 1. A follow-up survey found some improvement; organic chemicals were detected in only 17 percent of the samples (Bureau of Public Water Supply Protection 1991).

4. Bureau of Drinking Water 1988: 3.

5. Trent and Felice 1990: 5, 8.

6. Perrier's Web site, www.perrierusa.com (link to "about us," then "our history"), boasts that Julius Caesar's troops drank water from Perrier's source spring and that Emperor Napoleon III in 1863 "signed a decree . . . that the spring water was a natural mineral water."

7. Lauro 2001: C1.

8. Ferrier 2001: 17.

9. Lauro 2001: C1.

10. *Beverage Industry Annual Manual* 1992–93: 37.

11. Trent and Felice 1990: 5.

12. International Bottled Water Association (IBWA), "U.S. Bottled Water Market Volume, Growth, Consumption 1976–1999," http://www .bottledwater.org/public/volume growth and consumption.htm (accessed 1/19/2001).

13. U.S. House 1991a. In preparation for the hearings, the staff of the Commerce Committee's subcommittee organized a workshop on bottled water issues, *Proceedings of the Bottled Water Workshop,* and the subcommittee chair requested a report from the General Accounting Office.

14. U.S. GAO 1991b: 1, 2.

15. U.S. House 1991a: 7.

16. U.S. GAO 1991b: 2.

17. U.S. House 1991a: 2.

18. U.S. House 1991b: 22–32.

19. Ibid.: 46.

20. Olson 1999: vii.

21. U.S. House 1991a: 104–5.

22. The benzene was discovered by accident by a North Carolina state laboratory that "was using Perrier as a quality control sample to ensure the accuracy of its testing equipment." U.S. GAO 1991b: 6.

23. U.S. House 1991a: 3.

24. Ibid.: 29, 50. At IBWA's 1994 convention, the association's president, Colleen Porterfield, declared it was the industry's "dream . . . that the day will come when bottled water is the most regulated of drinking water" (Sfiligoy 1994: 3). More than a decade later, it is not clear that that day has come, in part because in spite of IBWA's fervently stated hopes, when stronger regulations are proposed, IBWA opposes them. In 1991, when a House bill proposed to shift regulation of bottled water from the FDA to the EPA, IBWA solicited contributions from member companies so it could mount a campaign to defeat the bill (U.S. House 1991a: 20). Later, IBWA would oppose a proposal that would have provided consumers more information about the contents of bottled waters. See "IBWA Comments on FDA Draft Study Report," May 1, 2000, www.bottledwaterweb.com/news/nw_050100.html (accessed July 8, 2006).

25. For details, see U.S. House 1991b: 64–71.

26. The product could also deteriorate after it was bottled, between the time it left the plant and the time it was consumed. Material could leach from the plastic bottle; bacteria could multiply as the water sat in the store or at home, sometimes for weeks, at room temperature. Ibid.: 72–94.

27. Ibid.: 64–94. The FDA did a survey of bottled waters in 1990. It

reported to Congress that almost a third of the samples "exceed[ed] testing standards for microbiological elements" (U.S. House 1991a: 9).

28. U.S. House 1991a: 185–86. Literature reviews that pointed to similar conclusions were submitted by the American Water Works Association and the Long Island Water Conference; U.S. House 1991b: 44–63.

29. U.S. House 1991a: 185–86.

30. Olson 1999: viii.

31. U.S. GAO 1991a: 10. See also NSF International, "Home Water Treatment Devices," www.nsf.org/consumer/drinking_water/dw_treatment.asp?program=WaterTre (accessed February 16, 2005).

32. U.S. GAO 1991a: 33.

33. Ibid.: 37, 16.

34. Gorman 1996.

35. U.S. GAO 1991a: 23.

36. Ibid.: 22.

37. www.HealthisWealthMaui.com, "Water Filter and Water Purifiers Information Page," www.healthiswealthmaui.com/MP%20MoreInfo.html#WhyNeed (accessed February 16, 2005).

38. E-mail correspondence with NSF staff member Greta Houlahan, February 24–25, 2005. NSF International's Web site, "NSF Products and Services Listings," lists seventy-six manufacturers that market certified carbon filter units and seventy-two that market certified reverse osmosis units; www.nsf.org/Certified/Common/Company.asp?submit4=All (accessed February 24, 2005).

39. NSF International, "Contamination Testing Protocols," www.nsf.org/consumer/drinking_water/dw_contaminant_protocols.asp?program=WaterTre (accessed February 16, 2005).

40. NSF International, "Drinking Water Myths," www.nsf.org/consumer/drinking_water/ww_myth.asp?program=WaterTre (accessed February 16, 2005).

41. Burros 2002a.

42. Yoon 2003. And see Lu et al. 2006.

43. Kegley, Katten, and Moses 2003.

44. Environmental News Service 2003.

45. Pearce and Mackenzie 1999; Cooper 2003.

46. Some large food corporations have tried to persuade federal regulators to interpret those standards in ways that allow them to continue to grow crops or to raise farm animals in conventional ways and still qualify for "organic" designation. The organic community thinks these interpretations would significantly weaken organic standards. Burros 2000b, 2003; Griscom 2004; Warner 2005.

47. Daley 2005.

48. Green Guide Editors, Grist Magazine, "Make Your Spring Cleaning Green," March 27, 2003, http://www.alternet.org/story/15486/ (accessed July 12, 2006).

49. Consumer Union, "The Consumer Union Guide to Environmental Labels," 2002, www.eco-labels.org/home.cfm (accessed July 12, 2006).

50. Consumer Union, "The Consumer Union Guide to Environmental Labels," 2002, www.eco-labels.org/agency.cfm?AgencyID=4 (accessed July 12, 2006).

51. Stehlin 1991.

52. Vincent 2005.

53. Merrill 2005: E3.

54. Ibid.: E1, E3.

55. Compliance with NSF's new standard for shower filters, NSF/ANSI 177, only means the filter removes free chlorine, which consumers can smell, not chlorination by-products, which are the real hazards. That is why it is considered an aesthetic standard, not a health standard. The distinction was confirmed via e-mail by Richard Andrew, Technical Manager of NSF's Drinking Water Treatment Program, February 7, 2005.

56. Using data from 1994 to 1996, Jacobs et al. (2000) found that 16 percent of Americans regularly use bottled water instead of tap water for drinking; 7 percent use bottled water instead of tap water for both drinking and cooking. Taking into account that aggregate consumption has gone up since then, we should assume that those numbers are somewhat higher today.

57. Water Quality Association, "Eighty-Six Percent of Americans Have Concerns about Their Home Drinking Water, New WQA Survey Finds," press release, April 23, 2001.

58. Yankelovich Partners 2000.

59. Ershow and Cantor 1989.

60. The breakdown of "intrinsic" water into the two different categories, "biological water" and "commercial water," is from Jacobs et al. 2000.

61. Groth, Benbrook, and Lutz 1999: 3, 16.

62. Schneider and Hill 2005.

63. National Cancer Institute 1997; Centers for Disease Control and Prevention and the National Cancer Institute 2001.

64. First reported by Dewailly et al. 1989. Studies done subsequently confirmed the transport of persistent organic pollutants and other harmful substances from the industrialized nations to the Arctic. See summaries of findings published by the Arctic Monitoring and Assessment Program (2002a, 2002b). See also the speech by Sheila Watt-Cloutier, president of the Inuit Circumpolar Conference, delivered at the Briefing on Persistent Organic Pollutants at the United Nations, New York, January 25,

2001, http://www.inuitcircumpolar.com/index.php?ID=99&Lang=En (accessed July 13, 2006).

65. Bridges 2002; Ebbert 2004; Environmental News Service 2002a, 2004, 2005d.

66. Bridges 2002.

67. Environmental News Service 2001a.

68. Quoted in Bridges 2002.

69. Ebbert 2004.

7. Political Anesthesia

1. Thomas and Thomas 1928: 572. The phrase begins one of Robert K. Merton's (1996) best-known essays, "The Self-Fulfilling Prophesy."

2. By the mid-1990s, two of the world's largest makers of PET bottles were each producing billions per year; Eastman Chemical Company, "45 Percent PET Capacity Increase," www.eastman.com/News_Center/ News_Archive/Corporate_News/1994/940712.asp (accessed April 19, 2002); Johnson Controls 1995. PET bottle use has grown substantially since then.

3. See Crittenden 1997 for one discussion of the energy and raw material inputs, as well as the waste stream outputs, from the manufacturing of different kinds of bottles.

4. In 2005, Eastman Chemical, which claims to be the world's largest producer of PET (polyethylene terephthalate, the type of resin popular for small-format bottles because it is lightweight, sturdy, and clear as glass), had the capacity to produce over 3 billion pounds of PET resin; Eastman Chemical 2005: 6. PET bottles are used for many foods and beverages, including large-volume items like Coke and Pepsi, not just water, but water is sold in containers made of other resins, such as HDPE, too. The 1.5 million ton estimate is by Ferrier 2001: 10.

5. American Plastics Council 2004: 1.

6. Environmental News Service 2002b.

7. Guthman 2004; Pollan 2006.

8. Griscom 2004; Warner 2005; Matthews 2006.

9. www.bottledwaterweb.com, "NRDC's Bottled Water Report: Pure Fact or Pure Fiction? An editorial from the BottledWaterWeb," www.bottledwaterweb.com/news/nw_041299.htm, quoting Olson 1999 (accessed August 1, 2000).

10. Ibid.

11. The $150 billion estimate comes from a 1999 U.S. EPA study (2001c); the $500 billion figure from U.S. EPA 2002; the trillion dollar estimate is reported by Revkin 2002.

12. Revkin 2002.

13. Bustillo and Weiss 2005.

14. In 2002, Congress considered a bill that would increase spending to $35 billion over five years; Revkin 2002. That begins to approach the low-end estimate for what is needed, but the bill stalled, and the fiscal situation has deteriorated considerably since, making passage of the bill very unlikely.

15. See Environmental News Service 2005c; Lillard 2005; 2006 polling results compiled on PollingReport.com, "Environment," http://www.pollingreport.com/enviro.htm (accessed July 15, 2006).

16. Lillard 2005. Similar results from another poll, again showing that Americans rank war, terrorism, jobs and economy, health care, and so on as more important issues than the "environment," are reported in *Rachel's Democracy & Health News*, "Environment Is Top Priority for Only 2% of Americans," #854, May 11, 2006, www.precaution.org/lib/06/prn_enviro_survey.060423.htm (accessed May 12, 2006).

17. Pollan 2001: 35.

18. Ibid.

19. Piketty and Saez 2001, figures 1 and 3.

20. Johnston 2005.

21. United for a Fair Economy, "Wealth Inequality Charts," www.faireconomy.org/research/wealth_charts.html (accessed October 4, 2005).

22. Szasz 1984, 1986b.

23. Szasz 1994.

24. McNeill 2000: 8, 15, 64, 121, 213, 360.

25. Hannah et al. 1994; Lazaroff 2001d; Postel, Daily, and Ehrlich 1996; Ramankutty and Foley 1999; Vitousek et al. 1986; Vitousek et al. 1997.

26. Intergovernmental Panel on Climate Change 2001d: 31.

27. Meadows et al. 1972: 23.

28. *Rachel's Environment & Health Weekly*, "Scientists Say Future Is in the Balance," #669, September 23, 1999, www.rachel.org/bulletin/bulletin.cfm?Issue_ID=1584 (accessed December 21, 2000).

29. The phrase is from Mann et al. 2003.

30. The Intergovernmental Panel on Climate Change has three working groups. Two of those groups' reports are cited here as 2001a, 2001b, 2001c, and 2001d.

31. Intergovernmental Panel on Climate Change 2001a: 2, 3, based on Mann, Bradley, and Hughes 1999.

32. Just of sample of recent work on the Arctic and Greenland: Arctic Climate Impact Assessment 2004; Overpeck et al. 2005; Rignot and Kanagaratnam 2006.

33. Again, just a sample of recent published studies: Parmesan and Yohe 2003; Root et al. 2003; Thomas et al. 2004; Parmesan and Galbraith 2004; Inkley et al. 2004.

34. Intergovernmental Panel on Climate Change 2001a: 13.

35. Stainforth et al. 2005.

36. Intergovernmental Panel on Climate Change 2001d: 43.

37. Intergovernmental Panel on Climate Change 2001d: 31. Higher water temperatures can also compromise water quality, already an issue in many places. Human populations will be further stressed by new diseases or changes in the ranges of existing diseases, as changes in climate push microbes and insects to new ranges; ibid.: 43.

38. Intergovernmental Panel on Climate Change 2001d: 70; 2001b: 16. That is not to say advanced, industrialized societies will find it easy to cope. A report by the National Assessment Synthesis Team (2000) predicts major impacts and great difficulties for the United States.

39. Environmental News Service 2001c; Agence France-Presse, "Refugees, Disease, Water and Food Shortages to Result from Global Warming," February 2, 2005, www.terradaily.com/2005/050202102932.i0omwo4d.html (accessed February 4, 2005); P. Brown 2005.

40. Environmental News Service 2002c.

41. Weart 2003a, 2003b; Adams, Maslin, and Thomas 1999.

42. Committee on Abrupt Climate Change 2002.

43. National Academies, "Publication Announcement: Possibility of Abrupt Climate Change Needs Research and Attention," December 11, 2001, www8.nationalacademies.org/onpinews/newsitem.aspx?Record ID=10136.

44. Intergovernmental Panel on Climate Change 2001d: 72.

45. McManus et al. 2004.

46. Intergovernmental Panel on Climate Change 2001d: 72.

47. Bryden, Longworth, and Cunningham 2005.

48. Scheffer et al. 2001.

49. National Academies, "Publication Announcement: Possibility of Abrupt Climate Change Needs Research and Attention." See also Weiss and Bradley 2001, and Jared Diamond's book-length study (2005) of how societies' problems extracting a living from nature lead to societal collapse.

50. Schwartz and Randall 2003: 14, 13, 14, 5, 2.

51. Ibid.: 16–17.

52. Environmental News Service 2005a.

53. Brechin 2003.

54. See Pew Research Center for the People and the Press 2006; Saad 2006; and PollingReport.com's summary of numerous other polls done in 2006.

55. Pew Research Center for the People and the Press 2006.

56. Responses to questions about willingness to pay gas tax and electricity tax on ABC News/Time/Stanford University Poll, March 2006, summarized on PollingReport.com.

57. PollingReport.com.

Conclusion

The title of the conclusion is an homage to Sigmund Freud's *The Future of an Illusion* (1927), Freud's analysis of religion as imaginary refuge. In the intervening eighty years, many of us have redirected our faith, from religion to the market, from God to the Commodity.

1. Schlesinger 1965: 910.

2. U.S. Department of Health and Human Services 2001, 2003, 2005.

3. TheOzoneHole, "Ozone Hole 2005," www.theOzonehole.com/Ozonehole2005.htm (accessed December 6, 2005); TheOzoneHole, "Ozone Hole History," www.theOzonehole.com/Ozoneholehistory.htm (accessed December 2, 2005).

4. CFCs were used in aerosol sprays until that use was banned in 1978. CFCs are solvents, so have also been used as cleaning agents.

5. NASA Advanced Supercomputing (NAS) Division, "Ozone Depletion, History and Politics," www.nas.nasa.gov/about/education/Ozone/history.html (accessed December 2, 2005).

6. For a good summary of the known health impacts of UV, see Longstreth et al. 1998.

7. Häder et al. 1998: 58.

8. A present-day example of what might have happened to us all: in Punta Arenas, at the tip of Chile, the southernmost significant human settlement, a city "under" the ozone hole when it reaches its greatest size during the Antarctic spring, the local weather report includes the level of UV for the day, and when UV levels reach a certain point, citizens are advised to stay indoors or apply high-SPF sunblock. Rohter 2002.

9. People tend to use too little of it, and they do not reapply it often enough. Neutrogena, for example, recommends applying a full ounce—about enough to fill a shot glass—every time one goes out, and reapplying "after prolonged swimming and towel drying." Besides the nontrivial fact that a four-ounce bottle of Neutrogena costs about $10.00 and thus applying Neutrogena once a day would cost a family over $900 per person per year, it is hard to imagine spreading *that much* sunscreen on one's body *every* time one applies it.

10. Controversy erupted in 1998 when a respected scientist, after reviewing available studies, argued that use of sunscreen did not reduce the incidence of skin cancer (Associated Press 1998). Although the dermatology establishment disapproved, her conclusions have held up pretty well in the intervening eight years (Johnson 2006). At the same time, adverse skin reactions to sunscreen are not rare (Nixon, Frowen, and Lewis 1997), and some sunscreens contain ingredients that are, themselves, considered carcinogenic (Rauch 2003).

11. George 2001; Holm-Hansen, Villafane, and Helbling 1997; Norris 1999; Schrope 2000.

12. Norris 1999: 520.

13. United Nations Environment Programme 1998: 2.

14. Häder et al. 1998.

15. Ibid.: 54.

16. Norris 1999: 520.

17. M. M. Caldwell et al. 1998: 40, 44.

18. Karentz and Bosch 2001: 3.

19. Häder et al. 1998: 63.

20. M. M. Caldwell et al. 1998: 44, 40.

21. A few years ago, full recovery was expected by 2040 or 2050. More recent studies push the expected date of recovery back by fifteen or twenty years (Davidson 2005; Environmental News Service 2006).

22. D. K. Caldwell 1999.

23. Browning 2003; Kuczynski 2001.

24. Porretto 2003.

25. P. L. Brown 1997. See also Rose 2001.

26. Riordan 2001.

27. Lloyd 2001.

28. Donn 2001.

29. Soon after 9/11 someone—the perpetrator has not yet been caught—sent some congressmen and senators letters filled with anthrax. The inverted quarantine reflex was triggered once again as Americans rushed to stock up on supplies of the antibiotic Cipro (Petersen 2001).

30. National Public Radio, "Building Safeguards," aired November 20, 2001.

31. Kristof 2002. On display at the Las Vegas Builders' Show were home safes specifically designed for a loaded gun because, as one exhibitor put it, "People are arming themselves like never before against home invasions"; McKee 2004.

32. Gibbs 2001: 40.

33. McKee 2004.

34. Strauss 2003.

35. For example, in *Montaillou: Cathars and Catholics in a French Village, 1294–1324* (1978), Emmanuel Le Roy Ladurie uses testimony coerced by the Inquisition to explore the characteristic ways peasants in that part of France perceived, thought about, and acted in their world. In other words, Le Roy Ladurie was attempting to understand their *mentalité*.

36. The Free Dictionary, "Mentality," www.thefreedictionary.com/mentality (accessed January 2006).

37. Everyday linguistic practice has us separate perception from cognition (assessment or evaluation of a situation), separate both from the development of an intention to take a particular course of action in response to that assessment, and separate all three from action. Inverted

quarantine cannot be neatly fitted into such a scheme, one that neatly orders the arc of human behavior from perception to action, cutting it into an orderly and linear sequence of conceptually separable phases.

Inverted quarantine is certainly an act, but as the discussion of organic food consumption, in chapter 7, showed it is not only that. Implicit in the distinction between "true naturals" and "health seekers" is the idea that intention matters. If you eat organic food because you are an activist committed to changing how society grows its food (in addition to doing it because it is personally healthier), you are not, strictly speaking, engaged in an act of inverted quarantine. If, on the other hand, you buy the exact same basket of goods only because you care about your health, that is inverted quarantine. So it is not just an act; motivation or intention determines the meaning of the act, determines if that act (otherwise exactly identical) is inverted quarantine or not.

Furthermore, one cannot separate intention from cognition, that is, the assessment or evaluation of a situation. The same threat (be that social or environmental) can be cognized either as nothing more than a threat to one's personal well-being, or it can be cognized as a collective, societal problem (which, of course, is also a threat to one's own well-being). How one characterizes a threat, how one explains to oneself what that threat is, what has caused it, and so on, is obviously related to the line of action one decides to take in response to it.

The phenomenologists long ago taught us that one cannot separate perception, even raw, "pure," sense perception, from cognition. It is never a stepwise, one-two process, first perception then interpretation. We actually organize our perceptions *as* they "enter" our senses. See Merleau-Ponty's *Phenomenology of Perception* (1962), and his *The Structure of Behavior* (1963). The distinction between true natural and health seeker does not begin to matter only at the moment intention forms; it is already present in the first moment of perception of threat. When the health seeker sees a bottle of milk from a dairy that injects its cows with bovine growth hormone, just to pick one example at random, she "sees" a product to avoid. When an environmental activist sees the same bottle of milk, she "sees" a system in desperate need of reform.

We need to have a concept that does not separate perception from interpretation, from intention and action, and, indeed, in social theory we can find a number of promising candidates. The term *mentalité* is one good candidate, as discussed in the text. Pierre Bourdieu's concept *habitus* tries to capture something quiet similar. Bourdieu offers many complementary, overlapping, redundant definitions of *habitus* in his *Outline of a Theory of Practice* (1977). Bourdieu tries to capture with this single term his understanding of the complex and subtle dialectic of culture and individual "practice." *Habitus* consists of "schemes of perception . . .

interpretation, . . . conception and action" (86, 80). Bourdieu also says *habitus* consists of "cognitive and motivating structures" (76). The schemes or structures of the *habitus* are "socially constituted" and "common to all members of the same group" (76, 86). These schemes are not hard and simple rules; they exist in the form of an "unconscious" but "durable . . . system of dispositions" (77, 72, 82). They act as the "generative principles" for "practices" (78, 72), "determin[ing for actors what is] 'reasonable' [versus] 'unreasonable' conduct," providing grounds for a "practical evaluation of the likelihood of the success of a given action in a given situation" (77).

If people have a marked tendency to react to threat not by joining with others to do something about the threat but by simply trying, individually, to barricade themselves from trouble, it is because that tendency expresses a certain characteristic way of "being-in-the-world"; that tendency is a manifestation of a particular *mentalité*, an expression in practice of the dominant *habitus*.

One more thing should be noted here: to identify an aspect of *mentalité* or a characteristic aspect of the *habitus* does not tell us where it comes from, what caused it. One could pursue that line of inquiry as well. One would have to start, I think, with an analysis of the development of what MacPherson dubbed "possessive individualism," and how that development is related to the culture of capitalism. Furthermore, the inverted quarantine *mentalité* does not just spontaneously "bubble up" from "below"; it is also encouraged "from above," in some cases (the atomic fallout shelter) by government officials, in more cases by commercial boosters and advertisers. It is both, really: as in the case of suburbanization, yes, it was boosterism and the selling of the new definition of the American Dream, but if it was manipulation, you would have to say it "went down" pretty easily. So a history of the development of this aspect of *mentalité* would require one to synthesize the power of ideological initiatives from above, with some theory about the mind-set or everyday culture of modern social life that recognizes the "readiness" "from below" to be responsive to such messages.

38. This was during the congressional proceedings that produced the Resource Conservation and Recovery Act (RCRA), the nation's flagship hazardous waste law. See Szasz 1986a.

39. For reviews of some of the literature on problems implementing regulatory laws, see Szasz 1982, chapter 2; Szasz 1994: 33–34.

40. See M. J. Cohen 2004. Cohen assesses only the first two years of the administration's environmental record and only a portion of that record (he discusses air pollution, climate change, and toxics remediation; he does not discuss other important issues, such as endangered species, resource extraction issues such as mining and logging, etc.). A quick

check of Web resources shows that environmental organizations, such as the Sierra Club, the Natural Resources Defense Council, Environmental Defense (formerly Environmental Defense Fund), all agree that the administration's policies rank it as probably the most anti-environmental administration in recent history.

41. One can say with a high degree of confidence, furthermore, that the practice of inverted quarantine will continue to exhibit a strong class character. In fact, given recent trends in income and wealth inequality (see chapter 7), that class gradient is only likely to get steeper. As conditions worsen, people of means will erect even taller, seemingly more impervious barriers between themselves and an increasingly chaotic and dangerous world. Further down the class hierarchy, people will be less and less able to barricade themselves. Even today most Americans cannot afford an all-organic diet and nothing but filtered or bottled water, and certainly cannot afford all-natural home furnishings and personal hygiene products, much less the sophisticated home security system, the safe room, the armored car, the private jet. While the wealthy retreat ever further into their personal safety bubbles, masses of people at the bottom will be exposed full force to chaotic, degraded social and environmental conditions. One can begin to imagine a future not unlike the one depicted in the movie *Blade Runner,* a society largely defined by class, essentially a two-class system, the elites self-quarantining themselves at the tops of skyscrapers, worlds apart from masses, who inhabit squalid, filthy, dangerous streets below.

References

Adams, Jonathan, Mark Maslin, and Ellen Thomas. 1999. "Sudden Climate Transitions during the Quaternary." *Progress in Physical Geography* 23, 1 (March): 1–36.

Alabama Cooperative Extension Service. 1995. "Drinking Water for Livestock and Poultry." wq-26.al, Auburn University, June, hermes.ecn .purdue.edu:8001/cgi/convertwq?7745 (accessed March 5, 2002).

American Lung Association. 2004. "State of the Air 2004." http:// lungaction.org/reports/stateoftheair2004.html (accessed May 25, 2006).

———. 2005. "State of the Air 2005." April 28, http://lungaction.org/ reports/stateoftheair2005.html (accessed May 25, 2006).

———. 2006. "State of the Air 2006." April 27, lungaction.org/reports/ stateoftheair2006.html (accessed May 25, 2006).

American Plastics Council. 2004. "2004 National Post-Consumer Plastics Recycling Report." www.plasticsresource.com/s_plasticsresource/ docs/1700/1646.pdf (accessed August 28, 2006).

Andelman, J. B., S. M. Meyers, and L. C. Wilder. 1986. "Volatilization of Organic Chemicals from Indoor Use of Water." In *Chemicals in the Environment*, ed. J. N. Lester, R. Perry, and R. M. Sterritt, 323–30. London: Selper.

Anderson, Elijah, and Douglas S. Massey, eds. 2001. *Problem of the Century: Racial Stratification in the United States*. New York: Russell Sage Foundation.

Arctic Climate Impact Assessment. 2004. *Impacts of a Warming Arctic*. Cambridge, U.K.: Cambridge University Press. Online at amap.no/ workdocs/index.cfm?dirsub=%2FACIA%2Foverview.

Arctic Monitoring and Assessment Program. 2002a. "Contaminants

Exposure of Arctic Humans and Biota." Fact Sheet, October, www
.amap.no/fs-exposure.pdf (accessed July 13, 2006).
———. 2002b. "Transport of Contaminants to the Arctic and Their
Fate." Fact Sheet, October, www.amap.no/fs-transport.pdf
(accessed July 13, 2006).
Associated Press. 1998. "Studies Doubt Sunscreens Stop a Cancer." *New
York Times*, February 18.
———. 2006. "Suspected Carcinogen Found in Cord Blood." *New York
Times*, February 6, www.truthout.org/issues_06/020706HA.shtml
(accessed February 7, 2006).
Atlas, Mark. 1995. "Inaccuracy in Pollutant Emission Data: TRI, TRI
Again." Unpublished manuscript.
Barboza, David. 2000. "Modified Foods Put Companies in a Quandary."
New York Times, June 4, 1, 25.
Barthes, Roland. 1972. *Mythologies*. New York: Hill and Wang.
Barzilay, Joshua I., Winkler G. Weinberg, and J. William Eley. 1999.
The Water We Drink: Water Quality and Its Effects on Health. New
Brunswick, N.J.: Rutgers University Press.
Baumgartner, M. P. 1988. *The Moral Order of a Suburb*. Oxford: Oxford
University Press.
Becker, Elizabeth. 2004. "Europe Approves Genetically Modified Corn
as Animal Feed." *New York Times*, July 20, query.nytimes.com/gst/
fullpage.html?sec=health&res=990DE2DE103AF933A15754C0A96
29C8B63 (accessed May 15, 2006).
Bell, Michelle L., Aidan McDermott, Scott L. Zeger, Jonathan M. Samet,
and Francesca Dominici. 2004. "Ozone and Short-term Mortality in
95 U.S. Urban Communities, 1987–2000." *Journal of the American
Medical Association* 292, 19 (November 17): 2372–78.
Benbrook, Charles M. 2004. "Minimizing Pesticide Dietary Exposure
through the Consumption of Organic Food: An Organic Center State
of Science Review." The Organic Center for Education & Promotion,
May, www.organic-center.org/reportfiles/PESTICIDE_SSR.pdf
(accessed April 26, 2006).
Berthold-Bond, Annie. 1999. *Better Basics for the Home: Simple Solutions for Less Toxic Living*. New York: Three Rivers Press.
Berube, Alan. 2003a. "Gaining but Losing Ground: Population Change
in Large Cities and Their Suburbs." In *Redefining Urban and Suburban America: Evidence from Census 2000*, vol. 1, ed. Bruce Katz and
Robert E. Lang, 33–50. Washington, D.C.: Brookings Institution Press.
———. 2003b. "Racial and Ethnic Change in the Nation's Largest Cities." In *Redefining Urban and Suburban America: Evidence from
Census 2000*, vol. 1, ed. Bruce Katz and Robert E. Lang, 137–53.
Washington, D.C.: Brookings Institution Press.

Berube, Alan, and William H. Frey. 2002. "A Decade of Mixed Blessings: Urban and Suburban Poverty in Census 2000." Washington, D.C.: Brookings Institution, Center on Urban & Metropolitan Policy, August.

Beverage Industry. 2000. "Bottle Water Soars." *Beverage Industry* 91, 9 (September): NP26–NP27.

Beverage Industry Annual Manual. 1986. "U.S. Liquid Consumption Trends," 10.

———. 1987. "Volume Up 13.5% in '85," 45–46.

———. 1988. "Growth Still Strong for Bottled Water Market," 47–50.

———. 1989. "Acquisitions Put Zest in Bottled Water Sales," 48–51.

———. 1989/90. "Bottled Water: An Era of Enviable, Unending Growth," 47–49.

———. 1992–93. "Consumer Confidence: Your Most Important Ingredient," 37–39.

Bier, Thomas. 2001. "Moving Up, Filtering Down: Metropolitan Housing Dynamics and Public Policy." Washington, D.C.: Brookings Institution, Center on Urban and Metropolitan Policy, September.

Blakely, Edward J., and Mary Gail Snyder. 1997. *Fortress America: Gated Communities in the United States.* Washington, D.C.: Brookings Institution Press.

Blisard, Noel, David Smallwood, and Steve Lutz. 1999. "Food Cost Indexes for Low-Income Households and the General Population." Washington, D.C.: USDA, Economic Research Service, Technical Bulletin no. 1872, February, www.ers.usda.gov/publications/tb1872/ tb1872.pdf and www.ers.usda.gov/publications/tb1872/tb1872a.pdf (accessed April 27, 2006).

Boddy, Trevor. 1992. "Underground and Overhead: Building the Analogous City." In *Variations on a Theme Park: The New American City and the End of Public Space,* ed. Michael Sorkin, 123–53. New York: Hill and Wang.

Bonczek, Richard, and Kenneth J. Markussen. 1987. "Survey of Volatile Organic Chemical Compounds in Bottled Water Products Distributed in New York State." Bureau of Public Water Supply Protection, New York State Department of Health, January.

Boorman, Gary A., Vicki Dellarco, June K. Dunnick, Robert E. Chapin, Sid Hunter, Fred Hauchman, Hank Gardner, Mike Cox, and Robert C. Sills. 1999. "Drinking Water Disinfection Byproducts: Review and Approach to Toxicity Evaluation." *Environmental Health Perspectives* 107, Su1 (February): 207–17.

Bourdieu, Pierre. 1977. *Outline of a Theory of Practice.* Trans. Richard Nice. Cambridge, Mass.: Cambridge University Press.

Boutrif, Ezzedine. 2000. "Risks of Undesired Substances in Feeds and

Animal Food Products." Proceedings of the International Workshop, Montpellier, France, December 11–13, wwww.cirad.fr/colloque/fao/pdf/24-boutrif.pdf (accessed May 12, 2006).

Bowers, Carol L. 1999. "The Quest for Water Purity." *Utility Business* 2, 8 (August): 52–54.

Boyer, Paul. 1994. *By the Bomb's Early Light: American Thought and Culture at the Dawn of the Atomic Age.* Chapel Hill: University of North Carolina Press.

Boyle, Hal. 1951. "Washington under the Bomb." *Collier's,* October 27, 20–21, special issue titled, "Preview of the War We Do Not Want."

Boyles, S. L., K. W. Wohlgemuth, G. R. Fisher, D. Lundstrom, and L. J. Johnson. 1988. "Livestock and Water." North Dakota State University Extension Service, AS-594.nd. hermes.ecn.purdue.edu:8001/cgi/convertwq?5599 (accessed March 5, 2002).

Brace, Charles Loring. 1967. *The Dangerous Classes of New York and Twenty Years' Work among Them.* Montclair, N.J.: Patterson Smith; reprinted from the 3rd ed., New York: Wynkoop & Hallenbeck, Publishers, 1880.

Brechin, Steve. 2003. "Comparative Public Opinion and Knowledge on Global Climatic Change and the Kyoto Protocol: The U.S. versus the World?" *International Journal of Sociology and Social Policy* 23: 106–34.

Brennan, John, and Edward W. Hill. 1999. "Where Are the Jobs?: Cities, Suburbs, and the Competition for Employment." Brookings Institution, Center on Urban and Metropolitan Policy, Survey Series, November.

Bridges, Andrew. 2002. "From Arsenic to Phosphorous, Tiny Pollutants Have Global Reach." Associated Press, *Santa Cruz Sentinel,* May 26, B-10.

Brody, Jane E. 2000. "Gene Altered Foods: A Case against Panic." *New York Times,* December 5, D8.

Brook, Robert D., Jeffrey R. Brook, Bruce Urch, Renaud Vincent, Sanjay Rajagopalan, and Frances Silverman. 2002. "Inhalation of Fine Particulate Air Pollution and Ozone Causes Acute Arterial Vasoconstriction in Healthy Adults." *Circulation* 105 (April 2): 1534–36.

Brookings Institution. n.d. Metropolitan Policy Program, Living Cities Databook Series, Living Cities Interactive Databooks, http://www.brookings.edu/es/urban/issues/demographics/demographics.htm and http://apps89.brookings.edu:89/livingcities/ (accessed July 29, 2004).

Brooks, David. 2002. "For Democrats, Time to Meet the Exurban Voter." *New York Times,* November 10, Week in Review, 3.

———. 2004. "Take a Ride to Exurbia." *New York Times,* November 9, national ed., A23.

Brown, JoAnne. 1988. "'A Is for Atom, B Is for Bomb': Civil Defense in

American Public Education, 1948–1963." *Journal of American History* 75, 1: 68–90.

Brown, Patricia Leigh. 1997. "The New 'God Forbid' Room." *New York Times,* September 25, B1.

Brown, Paul. 2005. "Climate Conference Hears Degree of Danger." *Guardian,* February 3, www.guardian.co.uk/climatechange/story/ 0,12374,1404453,00.html (accessed February 4, 2005).

Browning, Lynnley. 2003. "In Complicated Skies, Jet Services Gain Members." *New York Times,* April 20, Business, 4.

Bryden, Harry L., Hannah R. Longworth, and Stuart A. Cunningham. 2005. "Slowing of the Atlantic Meridional Overturning Circulation at 25° N." *Nature* 438 (December 1): 655–57.

Buckley, Brian. 2004. "Eat, Drink, and Be Wary: Chemicals Often Linger in Water after Treatment, *Poughkeepsie Journal,* January 18, www .mindfully.org/Water/2004/Chemicals-Water-Treatment18jan04.htm (accessed July 23, 2006).

Bureau of Drinking Water, Suffolk County Department of Health Services. 1988. "Water Quality Survey of Bottled Water and Bottled Water Substitutes," January.

Bureau of Public Water Supply Protection, New York State Department of Health. 1991. "1990 Survey of Bottled Water Products Distributed in New York State," August.

Burros, Marian. 1997. "U.S. to Subject Organic Foods, Long Ignored, to Federal Rules." *New York Times,* December 15, A1, A10.

———. 2002a. "Study Finds Far Less Pesticide Residue on Organic Produce." *New York Times,* May 8, A25.

———. 2002b. "A Definition at Last, but What Does It All Mean?" *New York Times,* October 16, D5.

———. 2002c. "What the Label Can Say, and What It Can't." *New York Times,* October 16, D5.

———. 2003. "U.S.D.A. Enters Debate on Organic Label Law." *New York Times,* February 26, D1, D5.

Busboom, Jan R., and Karen P. Penner. 2002. "Hormones and Meat." www.inform.umd.edu/EdRes/Topic/AgrEnv/ndd/safefood/ HORMONES_AND_MEAT.html (accessed March 9, 2002).

Bustillo, Miguel, and Kenneth R. Weiss. 2005. "Bush Plan Could Drain Effort to Clean Up Waters." *Los Angeles Times,* February 9.

Caldwell, Deborah Kovach. 1999. "Date of Reckoning: Christian Mothers, Others Spread Word on Y2K." *San Jose Mercury News,* January 30, E1.

Caldwell, M. M., L. O. Björn, J. F. Bornman, S. D. Flint, G. Kulandaivelu, A. H. Teramura, and M. Tevini. 1998. "Effects of Increased Solar Ultraviolet Radiation on Terrestrial Ecosystems." (United Nations

Environment Programme, "Environmental Effects of Ozone Depletion, 1998 Assessment," chapter 3). *Journal of Photochemistry and
Photobiology B: Biology* 46: 40–52. www.gcrio.org/ozone/chapter3
.pdf (accessed July 2, 2006).

California Legislature, Assembly Office of Research. 1985. "Bottled Water
and Vended Water: Are Consumers Getting Their Money's Worth?"
Report 061-A. Sacramento: Joint Publication Office. March.

Carpenter, David O., Kathleen Arcaro, and David C. Spink. 2002.
"Understanding the Human Health Effects of Chemical Mixtures."
Environmental Health Perspectives 110 (suppl. 1, February): 25–42.

Carroll, Malcolm. 2003. "Christian Ecology Link GM Crops Briefing
Paper." Christian Ecology Link, June, www.christian-ecology.org
.uk/gm-crops-bp.rtf (accessed May 15, 2006).

Centers for Disease Control and Prevention. 2000. "Surveillance for
Waterborne-Disease Outbreaks—United States, 1997–1998." *Mor-
bidity and Mortality Weekly Report,* May 26, vol. 49/No. SS-4.

Centers for Disease Control and Prevention and the National Cancer
Institute. 2001. "A Feasibility Study of the Health Consequences to
the American Population from Nuclear Weapons Tests Conducted by
the United States and Other Nations, Volume 1, Technical Report."
August. www.cdc.gov/nceh/radiation/fallout/falloutreport.pdf (accessed July 14, 2006).

Chamberlain, Owen. 1966. "The Effect of Civil Defense on Strategic
Planning." In *Civil Defense,* ed. Henry Eyring, 27–31. Washington,
D.C.: American Association for the Advancement of Science.

Chaney, Rufus L., James A. Ryan, Philip G. Reeves, and Robert W. Simmons. 2003. "Cadmium Risk Perception and Assessment Principles
and Procedures." www.icsu-scope.org/chmeeting/2003meeting/
abs_Chaney_etal.htm (accessed April 21, 2006).

Charles, Camille Zubrinsky. 2001. "Socioeconomic Status and Segregation: African Americans, Hispanics, and Asians in Los Angeles." In
Problem of the Century: Racial Stratification in the United States,
ed. Elijah Anderson and Douglas S. Massey, 271–89. New York:
Russell Sage Foundation.

Chevalier, Louis. 1973. *Laboring Classes and Dangerous Classes in Paris
during the First Half of the Nineteenth Century.* New York: Howard
Fertig.

Chui, Glennda. 2000. "Organic Going Mainstream." *San Jose Mercury
News,* March 29, 1E, 4E.

Cieslewicz, David J. 2002. "The Environmental Impacts of Sprawl."
In *Urban Sprawl: Causes, Consequences and Policy Responses,* ed.
Gregory D. Squires, 23–38. Washington, D.C.: Urban Institute Press.

Civil Defense Letter Committee. 1961. "Open Letter to John F. Kennedy."

Originally published in *New York Times,* November 10; reprinted in the *Bulletin of Atomic Scientists* 18, 2: 28–29.

Cohen, Adam. 2004. "The Supreme Struggle." *New York Times,* Education Life, January 18, 22–24, 38.

Cohen, Maurie J. 2004. "George W. Bush and the Environmental Protection Agency: A Midterm Appraisal." *Society and Natural Resources* 17: 69–88.

Commission of the European Communities. 2000. "Proposal for a Directive of the European Parliament and of the Council Amending Council Directive 96/22/EC Concerning the Prohibition on the Use in Stockfarming of Certain Substances Having a Hormonal or Thyrostatic Action and of Beta-agonists." Brussels: Commission of the European Communities, COM(2000) 320 final, 2000/0132 (COD), May 24, europa.eu.int/comm/food/fs/him/him01_en.pdf (accessed March 9, 2002).

Committee on Abrupt Climate Change, National Research Council. 2002. *Abrupt Climate Change: Inevitable Surprises.* Washington, D.C.: National Academies Press.

Commoner, Barry. 1966. "Feasibility of Biological Recovery from Nuclear Attack." In *Civil Defense,* ed. Henry Eyring, 89–109. Washington, D.C.: American Association for the Advancement of Science.

Commonweal Biomonitoring Resource Center. 2005. "Taking It All In: Documenting Chemical Pollution in Californians through Biomonitoring." Commonweal, Bolinas, Calif. www.commonweal.org/programs/download/TIAI_Lo-Rez.pdf (accessed April 13, 2006).

Connelly, Marjorie. 2004. "How Americans Voted: A Political Portrait." *New York Times,* November 7, Week in Review, 4.

Cooper, Audrey. 2003. "Closer Look at Pesticides in S.J. Rain." *Stockton Record,* November 10.

Costner, Pat, Beverly Thorpe, and Alexandra McPherson. 2005. "Sick of Dust; Chemicals in Common Products—A Needless Health Risk in Our Homes." Cleaner Production Action, Safer Products Project, March, www.safer-products.org/downloads/Dust Report.pdf (accessed January 10, 2006).

Crittenden, B. 1997. "Environmental Life Cycle Analysis: A Tool for Waste Minimisation." Institution of Chemical Engineers, *Environment97,* www.environment97.org/text/reception/r/techpapers/papers/g12.htm (accessed July 11, 2001).

Daley, Beth. 2005. "Eco-products in Demand, but Labels Can Be Murky." *Boston Globe,* February 9, A1.

Damron, B. L. 2002. "Water for Poultry." Fact Sheet AN 125, February, Animal Sciences Department, Florida Cooperative Extension Service,

Institute of Food and Agricultural Sciences, University of Florida, edis.ifas.ufl.edu/AN125 (accessed April 21, 2006).

Davidson, Keay. 2005. "Ozone Layer Recovery to Take Extra 15 Years: Scientists Blame Use of Banned Chemicals." SFGate.com, December 7, www.sfgate.com/cgi-bin/article.cgi?f=/c/a/2005/12/07/MNGSTG409K1.DTL (accessed December 7, 2005).

Davis, Mike. 1992. "Fortress Los Angeles: The Militarization of Urban Space." In *Variations on a Theme Park: The New American City and the End of Public Space,* ed. Michael Sorkin, 154–80. New York: Hill and Wang.

DeArmond, Michelle. 1996. "House-in-a-Bottle: Underground Homes Offer the Ultimate in Security." *Santa Cruz Sentinel,* May 5, E-1.

Dentler, Robert A., and Phillips Cutright. 1964. "Social Effects of Nuclear War." In *The New Sociology: Essays in Social Science and Social Theory in Honor of C. Wright Mills,* ed. Irving Louis Horowitz, 409–26. New York: Oxford University Press.

Denton, Nancy A. 1994. "Are African Americans Still Hypersegregated?" In *Residential Apartheid: The American Legacy,* ed. Robert D. Bullard, J. Eugene Grigsby III, and Charles Lee, 49–81. Los Angeles: UCLA Center for Afro-American Studies.

DeVries, Tom, and Anne Dorfman. 1980. "Surviving the End of the World." *New West 5,* 4 (February 25): 17–23.

Dewailly, Eric, Albert Nantel, Jean-Paul Weber, and François Meyer. 1989. "High Levels of PCBs in Breast Milk of Inuit Women from Arctic Quebec." *Bulletin of Environmental Contamination and Toxicology 43,* 5 (November): 641–46.

Diamand, Emily, Adrian Bebb, and Pete Riley. 1999. "Genetically Modified Animal Feeds." Friends of the Earth, September, www.foe.co.uk/resource/briefings/gm_animal_feed.pdf (accessed May 15, 2006).

Diamond, Jared. 2005. *Collapse: How Societies Choose to Fail or Succeed.* New York: Viking Penguin.

Diesenhouse, Susan. 1996. "In South Florida, Security Sells Houses." *New York Times,* March 3, national ed., 27.

Dillon, David. 1994. "Fortress America: More and More of Us Are Living behind Locked Gates." *Planning 60,* 6 (June): 8–14.

Donn, Jeff. 2001. "Fallout Shelters Resurface." *Santa Cruz Sentinel,* December 8, 1.

Dowling, John, and Evans M. Harrell, eds. 1987. *Civil Defense: A Choice of Disasters.* New York: American Institute of Physics.

Duany, Andres, Elizabeth Plater-Zyberk, and Jeff Speck. 2000. *Suburban Nation: The Rise of Sprawl and the Decline of the American Dream.* New York: North Point Press.

Dupuis, E. Melanie. 2000. "Not in My Body: rBGH and the Rise of Organic Milk." *Agriculture and Human Values 17:* 285–95.

Durham, Sharon. 2004. "Dainty Plant Outpowers Cadmium-Contaminated Soils." www.ars.usda.gov/is/AR/archive/sep04/plant0904.htm (accessed April 21, 2006).

Eastman Chemical Company. 2005. "2005 Data Book." library .corporate-ir.net/library/61/611/61107/items/193566/DataBook_05.pdf (accessed August 28, 2006).

Ebbert, Stephanie. 2004. "Asian Grit Aloft in New England; Pollutants Found to Travel Globally." *Boston Globe*, August 9. www.boston .com/news/local/new_hampshire/articles/2004/08/09/asian_grit_aloft_in_new_england/ (accessed August 10, 2004).

Egan, Timothy. 1995. "Many Seek Security in Private Communities." *New York Times*, September 3, national ed., 1.

Ehrlich, Anne H. 1988. "Nuclear Winter: Is Rehabilitation Possible?" In *Rehabilitating Damaged Ecosystems, Volume II*, ed. John Cairns Jr., 123–42. Boca Raton, La.: CRC Press.

Ehrlich, Paul. 1983. "Long-Term Biological Consequences of Nuclear War." *Science* 222, 4630 (December 23): 1293–1300.

———. 1984. "When Light Is Put Away: Ecological Effects of Nuclear War." In *The Counterfeit Ark: Crisis Relocation for Nuclear War*, ed. Jennifer Leaning and Langley Keyes, 247–71. Cambridge, Mass.: Ballinger Publishing Co.

Eichenseher, Tasha. 2004. "Is Your Furniture Making You Sick?" *Santa Cruz Sentinel*, January 22, C1–C2.

Ellin, Nan, ed. 1997. *Architecture of Fear*. New York: Princeton Architectural Press.

El Nasser, Haya. 2002. "Gated Communities More Popular, and Not Just for the Rich." *USA Today*, December 15. www.usatoday.com/news/nation/2002-12-15-gated-usat_x.htm (accessed March 31, 2003).

Environmental News Service. 2001a. "African Dust Clouds Feed Toxic Algae Blooms." August 29, ens-newswire.com/ens/aug2001/2001-08-29-06.asp (accessed August 30, 2001).

———. 2001b. "Arsenic Water Standard Challenged by Wood Preservers." Environmental News Service AmeriScan, March 2, ens-news .com/ens/mar2001/2001L-03-02-09.html (accessed March 5, 2001).

———. 2001c. "Running on Empty, a Report on World's Water Woes." March 22, ens.lycos.com/ens/mar2001/2001L-03-22-11.html (accessed March 23, 2001).

———. 2002a. "Mercury from China Rains Down on California." December 20, ens-news.com/ens/dec2002/2002-12-20-09.asp (accessed December 20, 2002).

———. 2002b. "Recycling Beverage Containers Saves Materials, Money." January 17, ens-news.com/ens/jan2002/2002L-01-17-09.html (accessed March 5, 2002).

————. 2002c. "World Security Depends on Averting Water Wars."
March 22, ens-news/ens/mar2002/2002L-03-22-1.html (accessed
March 23, 2002).

————. 2003. "Organic Farms Contaminated by Transgenic Organisms."
May 15, ens-news.com/ens/may2003/2003-05-15-09.asp (accessed
May 19, 2003).

————. 2004. "NASA Satellites Track Asia-Atlantic Smog Train." May 5,
http://www.ens-newswire.com/ens/may2004/2004-05-05-097.asp
(accessed May 5, 2004).

————. 2005a. "Joint Science Academies' Global Response to
Climate Change." June 7, www.ens-newswire.com/ens/jun2005/
2005-06-07-insaca.asp (accessed June 8, 2005).

————. 2005b. "Perchlorate in Breast Milk Found Nationwide." February 24, www.ens-newswire.com/ens/feb2005/2005-02-24-09.asp,
(accessed February 25, 2005).

————. 2005c. "Poll: Majority Wants U.S. Federal Trust Fund for
Clean Water." March 7, www.ens-newswire.com/ens/mar2005/
2005-03-07-02.asp (accessed March 8, 2005).

————. 2005d. "Silver from Asian Coal Burning Pollutes North Pacific."
March 14, www.ens-newswire.com/ens/mar2005/2005-03-14-05.asp
(accessed March 15, 2005).

————. 2006. "South Pole Ozone Recovery 20 Years Later Than Expected." June, www.ens-newswire.com/ens/jun2006/2006-06-20-05
.asp (accessed July 3, 2006).

Environmental Research Foundation (Princeton, N.J.). 1987. "EPA Says
All Landfills Leak, Even Those Using Best Available Liners." *Rachel's
Hazardous Waste News* 37, August 10.

————. 1988. "Leachate from Municipal Dumps Has Same Toxicity as
Leachate from Hazardous Waste Dumps." *Rachel's Hazardous Waste
News* 90, August 15.

Environmental Working Group. 2003. "BodyBurden: The Pollution in
People." Mount Sinai School of Medicine and Commonweal, January.
www.ewg.org: 16080/reports/bodyburden/ (accessed April 13, 2006).

————. 2005. "Body Burden—The Pollution in Newborns." July, www
.ewg.org/reports/bodyburden2/ (accessed September 13, 2005).

————. 2006. "Skin Deep: News about the Safety of Popular Health
and Beauty Brands." June, www.ewg.org/reports/skindeep/ (accessed
June 12, 2006).

Ershow, A. G., and K. P. Cantor. 1989. "Total Water and Tapwater Intake in the United States: Population-Based Estimates of Quantities
and Sources." Bethesda, Md.: National Cancer Institute, Life Sciences
Research Office, Report#263-MD-810264.

European Commission. 2000. "Review of Specific Documents Relating
to the SCVPH Opinion of 30 April 99 on the Potential Risk to Human

Health from Hormone Residues in Bovine Meat and Meat Products." Health and Consumer Protection Directorate-General, May 3, europa .eu.int/comm/food/fs/sc/scv/out33_en.pdf (accessed March 9, 2002).

———. 2002. "Opinion of the Scientific Committee on Veterinary Measures Relating to Public Health on Review of Previous SCVPH Opinions of 30 April 1999 and 3 May 2000 on the Potential Risk to Human Health from Hormone Residues in Bovine Meat and Meat Products." Health and Consumer Protection Directorate-General, April 10, europa.eu.int/comm/food/fs/sc/scv/out50_en.pdf (accessed May 9, 2006).

European Union. 2003. "Regulation (EC) No 1829/2003 of the European Parliament and of the Council, of 22 September 2003, on Genetically Modified Food and Feed." http://europa.eu/eur-lex/pri/en/oj/dat/2003/ l_268/l_26820031018en00010023.pdf (accessed May 15, 2006).

———. 2006. "GMOs Authorized for Feed Use in the European Union in Accordance with Directives 90/220/EEC and 2001/18/EC." http://ec.europa.eu/comm/food/food/biotechnology/authorisation/ 2001-18-ec_authorised_en.pdf (accessed May 15, 2006).

Ewen, Stuart. 1976. *Captains of Consciousness: Advertising and the Social Roots of the Consumer Culture.* New York: McGraw-Hill.

Eyring, Henry, ed. 1966. *Civil Defense.* Washington, D.C.: American Association for the Advancement of Science.

Faries, F. C., Jr., John Sweeten, and John C. Reagor. 1991. "Water Quality: Its Relationship to Livestock." 1-2374.tx, June, Texas Agricultural Extension Service, Texas A&M University, hermes.ecn .purdue.edu: 8001/cgi/convertwq?6490 (accessed March 5, 2002).

Fasenfest, David, Jason Booza, and Kurt Metzger. 2004. "Living Together: A New Look at Racial and Ethnic Integration in Metropolitan Neighborhoods." Brookings Institution, Center for Urban and Metropolitan Policy, The Living Cities Census Series, April.

Federation of Atomic Scientists. 1962. "Civil Defense Shelter Statement." *Bulletin of Atomic Scientists* 18, 2: 25–27.

———. 1997. "Global Nuclear Stockpiles: 1945–1997." *Bulletin of the Atomic Scientists.* www.bullatomsci.org/issues/nukenotes/ nd97nukenotes.htm (accessed April 14, 1999).

Feld, Bernard T. 1962. "More Important Than Shelters." *Bulletin of Atomic Scientists* 18, 4: 8–11.

Ferrier, Catherine. 2001. "Bottled Water: Understanding a Social Phenomenon." Discussion Paper, Commissioned by the World Wildlife Fund.

Firestone, David. 1999. "Many See Their Future in County with a Past." *New York Times,* April 8, A18.

Fishman, Robert. 1987. *Bourgeois Utopias: The Rise and Fall of Suburbia.* New York: Basic Books.

Fitzsimons, Neal. 1968. "Brief History of American Civil Defense." In

Who Speaks for Civil Defense? ed. Eugene P. Wigner, 28–48. New York: Charles Scribner's Sons.

Flacks, Richard. 1988. *Making History: The American Left and the American Mind.* New York: Columbia University Press.

Flusty, Steven. 1997. "Building Paranoia." In *Architecture of Fear,* ed. Nan Ellin, 47–59. New York: Princeton Architectural Press.

Ford, Timothy E. 1999. "Microbiological Safety of Drinking Water: United States and Global Perspectives." *Environmental Health Perspectives* 107 (Su1): 191–206.

Foulke, Judith E. 1993. "FDA Reports on Pesticides in Foods." U.S. Food and Drug Administration, FDA Consumer, June, http://vm.cfsan.fda .gov/~lrd/pesticid.html (accessed March 1, 2002).

Fox, Chuck. 2001. "Arsenic and Old Laws." *New York Times,* March 22, A27.

Fox, Maggie. 2006. "Common Soap Antiseptic Found in Crop Fields." today.reuters.com/news/articlenews.aspx?type=domesticNews& storyid=2006-05-02T211530Z_01_N02307050_RTRUKOC_0_ US-ANTISEPTIC.xml (accessed May 3, 2006).

Frey, William H. 1979. "Central City White Flight: Racial and Non-racial Causes." *American Sociological Review* 44 (June): 425–48.

———. 2001. "Melting Pot Suburbs: A Census 2000 Study of Suburban Diversity." Brookings Institution, Center on Urban and Metropolitan Policy, Census 2000 Series, June.

———. 2002. "Three Americas: The Rising Significance of Regions." *Journal of the American Planning Association* 68, 4 (October): 349–55. www.milkeninstitute.org/publications/publications.taf?function=detail &ID=242&cat=Arts (accessed March 5, 2004).

———. 2003. "Melting Pot Suburbs: A Study of Suburban Diversity." In *Redefining Urban and Suburban America: Evidence from Census 2000, Volume One,* ed. Bruce Katz and Robert E. Lang, 155–79. Washington, D.C.: Brookings Institution Press.

Fromm, Erich, and Michael Maccoby. 1962. "A Debate on the Question of Civil Defense." *Commentary* 33, 1 (January): 11–23.

Furstenberg, Frank F., Jr. 2001. "The Fading Dream: Prospects for Marriage in the Inner City." In *Problem of the Century: Racial Stratification in the United States,* ed. Elijah Anderson and Douglas S. Massey, 224–46. New York: Russell Sage Foundation.

Gainsborough, Juliet F. 2001. *Fenced Off: The Suburbanization of American Politics.* Washington, D.C.: Georgetown University Press.

Gans, Herbert J. 1972. "The Future of the Suburbs." In *The End of Innocence: A Suburban Reader,* ed. Charles M. Harr. Glenview, Ill.: Scott, Foresman and Company.

Gardner, Karen. 1997. "Bottled-Water Demand Bubbling Up." *Restaurants USA,* March.

Garreau, Joel. 1991. *Edge City: Life on the New Frontier.* New York: Doubleday.

Gauderman, W. James, Edward Avol, Frank Gilliland, Hita Vora, Duncan Thomas, Kiros Berhane, Rob McConnell, Nino Kuenzli, Fred Lurmann, Edward Rappaport, Helene Margolis, David Bates, and John Peters. 2004. "The Effect of Air Pollution on Lung Development from 10 to 18 Years of Age." *New England Journal of Medicine* 351, 11: 1057–67.

Gayk, William F. 1991. "The Taxpayers' Revolt." In *Postsuburban California: The Transformation of Postwar Orange County since World War II,* ed. Rob Kling, Spencer Olin, and Mark Poster, 281–300. Berkeley: University of California Press.

George, Alison. 2001. "Back from the Brink: Plants Fight Off the Ravages of the Ozone Hole." *New Scientist* 171, 2307 (September 8): 12.

Gerstell, Richard. 1950. *How to Survive an Atomic Bomb.* New York: Bantam Books.

Gibbs, Nancy. 2001. "We Gather Together." *Time,* November 19, 28–41.

Gilliom, Robert J., Jack E. Barbash, Charles G. Crawford, Pixie A. Hamilton, Jeffrey D. Martin, Naomi Nakagaki, Lisa H. Nowell, Jonathan C. Scott, Paul E. Stackelberg, Gail P. Thelin, and David M. Wolock. 2006. "Pesticides in the Nation's Streams and Ground Water, 1992–2001." U.S. Geological Survey, revised March 15.

Glaeser, Edward L., Matthew Kahn, and Chenghuan Chu. 2001. "Job Sprawl: Employment Location in U.S. Metropolitan Areas." Brookings Institution, Center on Urban and Metropolitan Policy, Survey Series, May.

Glaeser, Edward L., and Jesse M. Shapiro. 2003. "City Growth: Which Places Grew and Why." In *Redefining Urban and Suburban America: Evidence from Census 2000, Volume One,* ed. Bruce Katz and Robert E. Lang, 13–32. Washington, D.C.: Brookings Institution Press.

Glaeser, Edward L., and Jacob L. Vigdor. 2003. "Racial Segregation: Promising News." In *Redefining Urban and Suburban America: Evidence from Census 2000, Volume One,* ed. Bruce Katz and Robert E. Lang, 211–34. Washington, D.C.: Brookings Institution Press.

Glanz, James. 2002. "Almost All in U.S. Have Been Exposed to Fallout, Study Finds." *New York Times,* March 1.

Glenn, Norval D. 1973. "Suburbanization in the United States since World War II." In *The Urbanization of the Suburbs,* ed. Louis H. Massoti and Jeffrey K. Hadden, 51–78. Urban Affairs Annual Reviews, vol. 7. Beverly Hills, Calif.: Sage.

Goldsmith, William W., and Edward J. Blakely. 1992. *Separate Societies: Poverty and Inequality in U.S. Cities.* Philadelphia: Temple University Press.

Goodall, Leonard. 1972. "Governing the Suburbs." In *The End of*

Innocence: A Suburban Reader, ed. Charles M. Harr, 138–48. Glenview, Ill.: Scott, Foresman and Company.

Gorman, Christine. 1996. "Do Water Filters Work?" *Time,* June 10, 70.

Gottdiener, Mark, and George Kephart. 1991. "The Multinucleated Metropolitan Region: A Comparative Analysis." In *Postsuburban California: The Transformation of Postwar Orange County since World War II,* ed. Rob Kling, Spencer Olin, and Mark Poster, 31–54. Berkeley: University of California Press.

Green, Frank. 2006. "Grocers Go Green." *San Diego Tribune,* February 26.

Griscom, Amanda. 2004. "Organic: Friend or Faux?" *Grist Magazine,* May 18, www.gristmagazine.com/muck/muck051804.asp?source=daily (accessed May 18, 2004).

Groth, Edward, III, Charles M. Benbrook, and Karen Lutz. 1999. "Do You Know What You're Eating? An Analysis of U.S. Government Data on Pesticide Residues in Foods." Consumer Union of United States, February, www.consumersunion.org/pdf/Do_You_Know.pdf (accessed March 1, 2002).

Guthman, Julie. 2004. *Agrarian Dreams: The Paradox of Organic Farming in California.* Berkeley: University of California Press.

Guyer, Paul Q. 1980. "Livestock Water Quality." g-79-467.ne, University of Nebraska–Lincoln, Institute of Agriculture and Natural Resources, Cooperative Extension, hermes.ecn.purdue.edu:8001/cgi/convertwq?6391 (accessed March 5, 2002).

Hadden, Jeffrey K., and Josef J. Barton. 1973. "An Image That Will Not Die: Thoughts on the History of Anti-urban Ideology." In *The Urbanization of the Suburbs,* ed. Louis H. Massoti and Jeffrey K. Hadden, 79–116. Urban Affairs Annual Reviews, vol. 7. Beverly Hills, Calif.: Sage.

Häder, D.-P., H. D. Kumar, R. C. Smith, and R. C. Worrest. 1998. "Effects on Aquatic Ecosystems." United Nations Environment Programme, "Environmental Effects of Ozone Depletion, 1998 Assessment," ch. 4. *Journal of Photochemistry and Photobiology B: Biology* 46: 53–68. www.gcrio.org/ozone/chapter4.pdf (accessed July 2, 2006).

Haitch, Richard. 1983. "Follow-up on the News; Hiding from War." *New York Times,* March 27, sect. 1, 45.

Hannah, Lee, David Lohse, Charles Hutchinson, John L. Carr, and Ali Lankerani. 1994. "A Preliminary Inventory of Human Disturbance of World Ecosystems." *Ambio* 23, 4-5: 246–50.

Harr, Charles M., ed. 1972. *The End of Innocence: A Suburban Reader.* Glenview, Ill.: Scott, Foresman and Company.

Harris, Richard. 1999. "The Making of American Suburbs, 1900–1950s: A Reconstruction." In *Changing Suburbs: Foundation, Form, and*

Function, ed. Richard Harris and Peter J. Larkham, 91–110. London: E & FN Spon.

Hayes, Tyrone B., Paola Case, Sarah Chui, Duc Chung, Cathryn Haeffele, Kelly Haston, Melissa Lee, Vien Phoung Mai, Youssra Marjuoa, John Parker, and Mable Tsui. 2006. "Pesticide Mixtures, Endocrine Disruption, and Amphibian Declines: Are We Underestimating the Impact?" *Environmental Health Perspectives* 114, Number S-1: 40–50.

Hersey, John. 1946. *Hiroshima.* New York: A. A. Knopf.

Hobbs, Frank, and Nicole Stoops. 2002. "Demographic Trends in the Twentieth Century." CENSR-4, Washington, D.C.: U.S. Census Bureau, November.

Holmes, Steven A. 1997. "Many Whites Leaving Suburbs for Rural Areas." *New York Times,* October 19, 17.

Holm-Hansen, O., V. E. Villafane, and E. W. Helbling. 1997. "Effects of Solar Ultraviolet Radiation on Primary Production in Antarctic Waters." In *Antarctic Communities: Species, Structure and Survival,* ed. Bruno Battaglia, Jose Valencia, and D. W. H. Walton, 375–80. Cambridge: Cambridge University Press.

Horovitz, Bruce. 2004. "Organic Food Trend Chips Out a Niche in Snack Food Aisle." *USA Today,* June 21.

Howes, Ruth H., and Robert Ehrlich. 1987. "Long-range Recovery from Nuclear War." In *Civil Defense: A Choice of Disasters,* ed. John Dowling and Evans M. Harrell, 139–51. New York: American Institute of Physics.

Inglis, David R. 1962. "Shelters and the Chance of War." *Bulletin of Atomic Scientists* 18, 4: 18–22.

Inkley, D. B., M. G. Anderson, A. R. Blaustein, V. R. Burkett, B. Felzer, B. Griffith, J. Price, and T. L. Root. 2004. "Global Climate Change and Wildlife in North America." Wildlife Society Technical Review 04-2. Bethesda, Md.: The Wildlife Society.

Intergovernmental Panel on Climate Change. 2001a. "Summary for Policymakers; A Report of Working Group I of the Intergovernmental Panel on Climate Change." www.ipcc.ch/pub/spm22-01.pdf.

———. 2001b. "Summary for Policymakers; Climate Change 2001: Impacts, Adaptation, and Vulnerability." www.ipcc.ch/pub/wg2SPMfinal.pdf.

———. 2001c. "Technical Summary; A Report Accepted by Working Group I of the IPCC but Not Approved in Detail." www.ipcc.ch/pub/wg1TARtechsum.pdf.

———. 2001d. "Technical Summary–Climate Change 2001: Impacts, Adaptation, and Vulnerability, A Report of the Working Group II of the Intergovernmental Panel on Climate Change." www.ipcc.ch/pub/wg2TARtechsum.pdf.

Ionics, Inc. 1995. "Ionics 1995 Annual Report." Watertown, Mass.: Ionics, Inc.

Jackson, Kenneth T. 1985. *Crabgrass Frontier: The Suburbanization of the United States,* New York: Oxford University Press.

Jacobs, Helen L., Julie T. Du, Henry D. Kahn, and Kathleen A. Stralka. 2000. "Estimated Per Capita Water Ingestion in the United States." Washington, D.C.: Environmental Protection Agency, EPA-822-R-00-008, April.

Jacobs, Jane. 1993 [1961]. *The Death and Life of Great American Cities.* New York: Modern Library.

Jargowsky, Paul A. 2003. "Stunning Progress, Hidden Problems: The Dramatic Decline of Concentrated Poverty in the 1990s." Brookings Institution, Center for Urban and Metropolitan Policy, The Living Cities Census Series, May.

Jehl, Douglas. 2001. "E.P.A. to Abandon New Arsenic Limits for Water Supply." *New York Times,* March 2, A1, A20.

Johnson Controls, Inc. 1995. "Annual Report 1995; Plastics Packaging Solutions." www.johnsoncontrols.com/annual_report/ar95/ar95_5.htm (accessed April 19, 2002).

Johnson, Linda A. 2006. "Sunscreens Faulted on Cancer Protection." *Santa Cruz Sentinel,* June 16.

Johnson, Robert L., Jr. 2004. "Relative Effects of Air Pollution on Lungs and Heart." *Circulation* 109 (January): 5–7.

Johnston, David C. 2005. "At the Very Top, a Surge in Income in '03." *New York Times,* October 5, C4.

Kahlenberg, Rebecca R. 2003. "Getting Clean and Green: Earth-Friendly Products Are Going Mainstream, One Shelf at a Time." *Washington Post,* September 18, H1.

Kahn, Herman. 1961. *On Thermonuclear War.* Princeton, N.J.: Princeton University Press.

———. 1962a. "A Debate on the Question of Civil Defense." *Commentary* 33, 1 (January): 1–11.

———. 1962b. *Thinking about the Unthinkable.* New York: Horizon Press.

Kaplan, Fred. 1983. *The Wizards of Armageddon.* New York: Simon and Schuster.

Karentz, Deneb, and Isidro Bosch. 2001. "Influence of Ozone-Related Increases in Ultraviolet Radiation on Antarctic Marine Organism." *American Zoologist* 41, 1 (February): 3–16.

Kasarda, John D., and George V. Redfearn. 1996. "Differential Patterns of City and Suburban Growth in the United States." In *American Cities: A Collection of Essays, Volume 1: Urbanization and the Growth of Cities,* ed. Neil L. Shumsky, 405–28. New York: Garland Publishing.

Katz, Arthur M. 1982. *Life after Nuclear War: The Economic and Social Impacts of Nuclear Attacks on the United States.* Cambridge, Mass.: Ballinger.

Katz, Bruce, and Jennifer Bradley. 1999. "Divided We Sprawl." *Atlantic Monthly* 284, 6 (December): 26–30, 38–42.

Katz, Bruce, and Robert E. Lang, eds. 2003. *Redefining Urban and Suburban America: Evidence from Census 2000, Volume One.* Washington, D.C.: Brookings Institution Press.

Kegley, Susan, Anne Katten, and Marion Moses. 2003. "Secondhand Pesticides: Airborne Pesticide Drift in California." Published jointly by Pesticide Action Network, California Rural Legal Assistance Foundation, Pesticide Education Center, and Californians for Pesticide Reform.

Kennedy, David J. 1995. "Residential Associations as State Actors: Regulating the Impact of Gated Communities on Nonmembers." *Yale Law Journal* 105, 3 (December): 761–93.

Kennedy, John F. 1961. "The Berlin Crisis: We Will Fulfill Our Pledge to West Germany." Speech delivered to the nation over television and radio, Washington, D.C., July 25. *Vital Speeches of the Day* 27, 21 (August 15): 642–45.

Kerr, Thomas J. 1983. *Civil Defense in the U.S.: Bandaid for a Holocaust?* Boulder, Colo.: Westview Press.

Kling, Rob, Spencer Olin, and Mark Poster, eds. 1991. *Postsuburban California: The Transformation of Postwar Orange County since World War II.* Berkeley: University of California Press.

Klonsky, Karen, and Catherine Greene. 2005. "Widespread Adoption of Organic Agriculture in the U.S.: Are Market-Driven Policies Enough?" May 16, www.organicaginfo.org/upload/GreeneC.AAEA%20organiccombinedfinal.5.16.05.pdf (accessed April 24, 2006).

Kolpin, Dana W., Edward T. Furlong, Michael T. Meyer, E. Michael Thurman, Steven D. Zaugg, Larry B. Barber, and Herbert T. Buxton. 2002. "Pharmaceuticals, Hormones, and Other Organic Wastewater Contaminants in U.S. Streams, 1999–2000: A National Reconnaissance." *Environmental Science and Technology* 36, 6: 1202–11.

Kortbech-Olesen, Rudy, and Tim Larsen. 2001. "The U.S. Market for Organic Fresh Produce." Port of Spain, Trinidad and Tobago: Conference on Supporting the Diversification of Exports in the Caribbean/Latin American Region through the Development of Organic Horticulture, October 8–10. www.intracen.org/mds/sectors/organic/usmarket.htm (accessed April 24, 2006).

Kristof, Nicholas D. 2002. "Chicks with Guns." *New York Times,* March 8, A23.

Kuczynski, Alex. 2001. "Private Skies of the Very Rich." *New York Times,* October 7, sect. 9, 1, 6.

La Ferla, Ruth. 2001. "Fashionistas, Ecofriendly and All-Natural." *New York Times*, July 15, sect. 9, 1.

Lang, Robert E. 2002. "Metropolitan Growth Counties." Fannie Mae Foundation, http://www.mi.vt.edu/Census2000/PDFfiles/Growth-Counties.pdf (accessed before September 2005).

———. 2004. Letter to the Editor. *New York Times*, November 12, national ed., A22.

Lang, Robert E., and Karen A. Danielsen. 1997. "Gated Communities in America: Walling Out the World?" *Housing Policy Debate* 8, 4: 867–99.

Lang, Robert E., and Patrick A. Simmons. 2003. "'Boomburbs': The Emergence of Large, Fast-Growing Suburban Cities." In *Redefining Urban and Suburban America: Evidence from Census 2000, Volume One*, ed. Bruce Katz and Robert E. Lang, 101–16. Washington, D.C.: Brookings Institution Press.

Lapp, Ralph E. 1954. "Civil Defense Faces New Perils." *Bulletin of the Atomic Scientists* 10 (November): 349–51. Reprinted in *The American Atom: A Documentary History of Nuclear Policies from the Discovery of Fission to the Present, 1939–1984*, ed. Robert C. Williams and Philip L. Cantelon, 183–90. Philadelphia: University of Pennsylvania Press, 1984.

Larson, Debra L. 2001. "No Significant Differences Found in Animals Fed GMO Corn and Soybeans." University of Illinois, College of ACES, April 25, www.aces.uiuc.edu/news/stories/news1374.html (accessed May 12, 2006).

Lauro, Patricia Winters. 2001. "Trying to Sparkle among the Hip; Perrier Hopes for a Comeback with Its New Ads." *New York Times*, June 21, C1, C6.

Lazaroff, Cat. 2001a. "Bush Administration Adopts Clinton's Arsenic Rule." Environmental News Service, November 1, ens-news.com/ens/nov2001/2001L-11-01-06.html (accessed November 2, 2001).

———. 2001b. "Study Tallies Americans' Exposure to Chemical Contaminants." Environmental News Service, March 22, ens.lycos.com/ens/mar2001/2001L-03-22-06.html (accessed March 22, 2001).

———. 2001c. "U.S. Moves to Limit Arsenic in Drinking Water." Environmental News Service, January 18, http://ens.lycos.com/ens/jan2001/2001L-01-18-06.html (accessed January 19, 2001).

———. 2001d. "World Land Database Charts Course of Human Consumption." Environmental News Service, July 11, enslycos.com/ens/jul2001/2001L-07-11-06.html (accessed July 12, 2001).

Leaning, Jennifer, and Langley Keyes, eds. 1984. *The Counterfeit Ark: Crisis Relocation for Nuclear War*. Cambridge, Mass.: Ballinger Publishing Co.

Le Roy Ladurie, Emmanuel. 1978. *Montaillou: Cathars and Catholics in a French Village, 1294–1324*. Trans. by Barbara Bray. London: Scolar.

Levi, Barbara G. 1987. "Civil Defense Implications of Nuclear Winter." In *Civil Defense: A Choice of Disasters*, ed. John Dowling and Evans M. Harrell, 125–37. New York: American Institute of Physics.

Lifton, Robert Jay, and Richard Falk. 1982. *Indefensible Weapons: The Political and Psychological Case against Nuclearism*. New York: BasicBooks/HarperCollins.

Lifton, Robert Jay, Eric Markusen, and Dorothy Austin. 1984. "The Second Death: Psychological Survival after Nuclear War." In *The Counterfeit Ark: Crisis Relocation for Nuclear War*, ed. Jennifer Leaning and Langley Keyes, 285–300. Cambridge, Mass.: Ballinger Publishing Co.

Lillard, Margaret. 2005. "Survey: Voters Care about Environment, but Not at Ballot Box." *Boston Globe*, September 20.

Littlefield, Joanne. 2004. "Measuring Perchlorate Levels in Lettuce." University of Arizona, January 28, uanews.org/cgi-bin/WebObjects/ UANews.woa/1/wa/SRStoryDetails?ArticleID=8562 (accessed February 4, 2004).

Lloyd, Caro. 2001. "Give Me Shelter! Will 9/11 Turn Even Our Peace-Loving Locals into Security-Hungry Survivalists?" SF Gate, October 23, www.sfgate.com/cgi-bin/article.cgi?file=/gate/archive/2001/ 10/23/carollloyd.dtl (accessed October 23, 2001).

Logan, John R. 2003. "Ethnic Diversity Grows, Neighborhood Integration Lags." In *Redefining Urban and Suburban America: Evidence from Census 2000, Volume One*, ed. Bruce Katz and Robert E. Lang, 235–55. Washington, D.C.: Brookings Institution Press.

Logan, John R., and Harvey L. Molotch. 1987. *Urban Fortunes: The Political Economy of Place*. Berkeley: University of California Press.

Longstreth, J., F. R. de Gruijl, M. L. Kripke, S. Abseck, F. Arnold, H. I. Slaper, G. Velders, Y. Takizawa, and J. C. van der Leun. 1998. "Health Risks." United Nations Environment Programme, "Environmental Effects of Ozone Depletion, 1998 Assessment," ch. 2. *Journal of Photochemistry and Photobiology B: Biology* 46: 20–39. www .gcrio.org/ozone/chapter2.pdf (accessed July 2, 2006).

Low, Setha. 2003. *Behind the Gates: Life, Security, and the Pursuit of Happiness in Fortress America*. New York: Routledge.

Lu, Chensheng, Kathryn Toepel, Rene Irish, Richard A. Fenske, Dana B. Barr, and Roberto Bravo. 2006. "Organic Diets Significantly Lower Children's Dietary Exposure to Organophosphorus Pesticides." *Environmental Health Perspectives* 114, 2 (February): 260–63.

Lucy, William H., and David L. Phillips. 2001. "Suburbs and the Census:

Patterns of Growth and Decline." Brookings Institution, Center on Urban and Metropolitan Policy, Survey Series, December.

———. 2003. "Suburbs: Patterns of Growth and Decline." In *Redefining Urban and Suburban America: Evidence from Census 2000, Volume One*, ed. Bruce Katz and Robert E. Lang, 117–36. Washington, D.C.: Brookings Institution Press.

Lunder, Sonya, and Renee Sharp. 2003. "Mothers' Milk: Record Levels of Toxic Fire Retardants Found in American Mothers' Breast Milk." www.ewg.org/reports_content/mothersmilk/pdf/mothersmilk_final .pdf (accessed April 14, 2006).

Lyman, Rick. 2006. "Surge of Population in the Exurbs Continues." *New York Times,* June 22, A10.

Madden, Janice F. 2001. "Do Racial Composition and Segregation Affect Economic Outcomes in Metropolitan Areas?" In *Problem of the Century: Racial Stratification in the United States,* ed. Elijah Anderson and Douglas S. Massey, 290–316. New York: Russell Sage Foundation.

Mann, Michael E., Caspar C. Amman, Ray S. Bradley, Keith R. Briffa, Philip D. Jones, Tim J. Osborn, Tom J. Crowley, Malcolm Hughes, Michael Oppenheimer, Jonathan T. Overpeck, Scott Rutherford, Kevin E. Trenberth, and Tom M. L. Wigley. 2003. "On Past Temperatures and Anomalous Late-Twentieth-Century Warmth." *EOS, Transactions, American Geophysical Union* 84, 27 (July 8): 256–57.

Mann, Michael E., Raymond S. Bradley, and Malcolm K. Hughes. 1999. "Northern Hemisphere Temperatures during the Past Millennium: Interferences, Uncertainties, and Limitations." *Geophysical Research Letters* 26, 6 (March 15): 759–62.

Martin, Thomas L., Jr., and Donald C. Latham. 1963. *Strategy for Survival.* Tucson: University of Arizona Press.

Massey, Douglas S. 2001. "Segregation and Violent Crime in Urban America." In *Problem of the Century: Racial Stratification in the United States,* ed. Elijah Anderson and Douglas S. Massey, 317–44. New York: Russell Sage Foundation.

Massey, Douglas S., and Nancy A. Denton. 1989. "Hypersegregation in U.S. Metropolitan Areas: Black and Hispanic Segregation along Five Dimensions." *Demography* 26: 373–91.

———. 1993. *American Apartheid: Segregation and the Making of the Underclass.* Cambridge, Mass.: Harvard University Press.

Massey, Rachel. 2001. "Arsenic from Your Tap." *Rachel's Environmental & Health News* 722, April 12, www.rachel.org/bulletin/index .cfm?issue_ID=1981 (accessed April 18, 2001).

Massoti, Louis H., and Jeffrey K. Hadden, eds. 1973. *The Urbanization of the Suburbs.* Urban Affairs Annual Reviews, vol. 7. Beverly Hills, Calif.: Sage.

Matthews, Kymberlie Adams. 2006. "The Rotten Side of Organics— Interview with Ronnie Cummins." *Satya,* April 15, posted on RACHEL, 855, May 18, www.precaution.org/lib/06/prn_ronnie_ cummins_interview.60401.htm (accessed May 22, 2006).

Maugh, Thomas H., II. 2003. "'Safe' Lead Levels Lower IQ in Children, Study Finds." *Los Angeles Times,* April 17.

McCarthy, James E. 1999. "97007: Clean Air Act Issues in the 105th Congress." March 3, ncseonline.org/NLE/CRSreports/Air/air-14.cfm (accessed May 29, 2003).

McGoey, Chris E. n.d. "Gated Community: Access Control Issues." http://www.crimedoctor.com/gated.htm (accessed March 31, 2003).

McHugh, L. C., S.J. 1961. "Ethics at the Shelter Doorway." *America* 105 (September 30): 824–26.

McKee, Bradford. 2004. "Fortress Home: Welcome Mat Bites." *New York Times,* January 22, D1.

McManus, J. F., R. Francois, J.-M. Gherardi, L. D. Keigwin, and S. Brown-Leger. 2004. "Collapse and Rapid Resumption of Atlantic Meridional Circulation Linked to Deglacial Climate Changes." *Nature* 428 (April 22): 834–37.

McNeill, John R. 2000. *Something New under the Sun: An Environmental History of the Twentieth-Century World.* New York: W. W. Norton & Company.

Meadows, Donella H., Dennis L. Meadows, Jorgen Randers, and William W. Behrens III. 1972. *The Limits to Growth: A Report from the Club of Rome's Project on the Predicament of Mankind.* New York: Universe Books.

Mellon, Margaret, Charles Benbrook, and Karen L. Benbrook. 2001. "Hogging It: Estimates of Antimicrobial Abuse in Livestock." Union of Concerned Scientists, January.

Melosi, Martin V., ed. 1980. *Pollution and Reform in American Cities, 1870–1930.* Austin: University of Texas Press.

———. 1981. *Garbage in the Cities: Refuse, Reform, and the Environment: 1880–1980.* College Station: Texas A&M University Press.

———. 2000. *The Sanitary City: Urban Infrastructure in America from Colonial Times to the Present.* Baltimore: Johns Hopkins University Press.

Merleau-Ponty, Maurice. 1962. *The Phenomenology of Perception.* Trans. Colin Smith. New York: Humanities Press.

———. 1963. *The Structure of Behavior.* Trans. Alden L. Fisher. Boston: Beacon Press.

Merrill, Jessica. 2005. "Is It Organic? Well, Maybe." *New York Times,* October 20, E1, E3.

Merton, Robert K. 1996 [1948]. "The Self-Fulfilling Prophesy." In *On Social Structure and Science,* ed. Piotr Sztompka, 183–201. Chicago: University of Chicago Press.

Meyer, K. B. 1990. "Water Quality for Animals." wq-9.in, September. hermes.ecn.purdue.edu: 8001/cgi/convertwq?5989 (accessed March 5, 2002).

Mitchell, Donald W. 1962. *Civil Defense: Planning for Survival and Recovery.* Washington, D.C.: Industrial College of the Armed Forces.

Muller, Peter O. 1996. "The Evolution of American Suburbs: A Geographical Interpretation." In *American Cities: A Collection of Essays, Volume 1: Urbanization and the Growth of Cities,* ed. Neil L. Shumsky, 395–404. New York: Garland Publishing.

Mumford, Lewis. 1961. *The City in History: Its Origins, Its Transformations, and Its Prospects.* New York: Harcourt Brace Jovanovich.

Murphy, Charles J. V. 1953. "A-Bomb vs. House." *Life* 34, 13 (March 30): 21–23.

Murphy, Eileen, Brian Buckley, Lee Lippincott, Ill Yang, and Bob Rosen. 2003. "The Characterization of Tentatively Identified Compounds (TICs) in Samples from Public Water Systems in New Jersey." New Jersey Department of Environmental Protection, Division of Science, Research & Technology, March.

Nafstad, P., L. L. Håheim, B. Oftedal, F. Gram, I. Holme, I. Hjermann, and P. Leren. 2003. "Lung Cancer and Air Pollution: A 27-Year Follow-up of 16,209 Norwegian Men." *Thorax* 58: 1071–76.

National Academies. 2001. "Publication Announcement: Possibility of Abrupt Climate Change Needs Research and Attention." December 11, www8.nationalacademies.org/onpinews/newsitem.aspx?RecordID=10136.

National Assessment Synthesis Team. 2000. "Climate Change Impacts on the United States: The Potential Consequences of Climate Variability and Change." U.S. Global Change Research Program, Washington, D.C., www.gcrio.org/NationalAssessment/index.htm.

National Cancer Institute. 1997. "Study Estimating Thyroid Doses of I-131 Received by Americans from Nevada Atmospheric Nuclear Bomb Test." http://rex.nci.nih.gov/massmedia/Fallout/contents.html (accessed July 14, 2006).

National Environmental Education and Training Foundation. 1999. "Report Card on Safe Drinking Water Attitudes, Knowledge and Behaviors." Opinion poll carried out for NEETF by Roper Starch Worldwide, July, www.neetf.org/pubs/watesummary.doc (accessed July 27, 2006).

Ness, Carol. 2006. "Green Giants: Mega-Producers Tip Scales as Organics Go Mainstream." *San Francisco Chronicle*, April 30. sfgate.com/cgi-bin/article.cgi?f=/c/a/2006/04/30/ORGANIC.TMP (accessed May 1, 2006).

New York State Civil Defense Commission. 195? "You and the Atomic Bomb: What to Do in Case of an Atomic Attack." Public Pamphlet 1.

Nixon, R. L., K. E. Frowen, and A. E. Lewis. 1997. "Skin Reactions to Sunscreens." *Australas Journal of Dermatology* 38 Suppl. 1: S83–85.

Norris, Scott. 1999. "Marine Life in the Limelight." *BioScience* 49, 7 (July): 520–26.

Nussbaum, Alex. 2003. "N.J. Water Contains Traces of Daily Life." *North Jersey News*, March 5, http://www.northjersey.com/page.php?qstr=eXJpcnk3ZjcxN2Y3dnFlZUVFeXkyJmZnYmVsN2Y3dnFlZUVFeXk2MzQ5NjQw (accessed March 6, 2003).

Oakes, Guy. 1994. *The Imaginary War: Civil Defense and American Cold War Culture*. New York: Oxford University Press.

Oliver, J. Eric. 2001. *Democracy in Suburbia*. Princeton, N.J.: Princeton University Press.

Olson, Erik. 2002. "What's on Tap? Grading Drinking Water in U.S. Cities, Early Release California Edition." Natural Resources Defense Council, October.

Olson, Erik D., with Diane Poling and Gina Solomon. 1999. *Bottled Water: Pure Drink or Pure Hype?* New York: Natural Resources Defense Council.

Organic Trade Association. 2004. "OTA Survey: U.S. Organic Sales Reach $10.8 Billion." *What's News in Organic* 28 (summer), www.ota.com/pics/documents/WhatsNews28.pdf (accessed April 26, 2006).

———. 2005a. "Consumer Profile Facts." www.ota.com/organic/mt/consumer.html (accessed April 24, 2006).

———. 2005b. "Industry Stats." www.organicexpo.com/ato06/custom/2005_site/stats.shtml (accessed April 21, 2006).

———. 2005c. "Organic Food Facts." www.ota.com/organic/mt/food.html (accessed April 26, 2006).

———. 2005d. "The Organic Industry." www.ota.com/pics/documents/The_Organic_Industry_Flyer.pdf (accessed April 21, 2006).

Overpeck, J. T., et al. 2005. "Arctic System on Trajectory to New, Seasonally Ice-Free State." *EOS, Transactions, American Geophysical Union* 86, 34 (August 23): 309–16.

Owens, John B. 1997. "Westec Story: Gated Communities and the Fourth Amendment." *American Criminal Law Review* 34, 3 (Spring): 1127–60.

Panofsky, Wolfgang K. H. 1966. "Civil Defense as Insurance and as

Military Strategy." In *Civil Defense,* ed. Henry Eyring, 11–25. Washington, D.C.: American Association for the Advancement of Science.

Parmesan, Camille, and Hector Galbraith. 2004. "Observed Impacts of Global Climate Change in the U.S." Pew Center on Global Climate Change, November.

Parmesan, Camille, and Gary Yohe. 2003. "A Globally Coherent Fingerprint of Climate Change Impacts across Natural Systems." *Nature* 421 (January 2): 37–42.

Parrot, Wayne. 2005. "GM Animal Feed Safety Papers (abstracts)." AgBioWorld, October, www.agbioworld.org/biotech-info/articles/agbio-articles/GMfeedsafetypapers.html (accessed May 12, 2006).

Peabody, Erin. 2004. "With Just a Sprinkle, Plants Soak Up More Selenium." USDA Agricultural Research Service, February 24, www.ars.usda.gov/is/pr/2004/040224.htm?pf=1 (accessed April 21, 2006).

Pearce, Fred, and Debora Mackenzie. 1999. "It's Raining Pesticides." *World Press Review,* July, 37.

Pegg, J. R. 2003. "Toxic Chemical Study Sounds Warning for Children." Environmental News Service, February 4, ens-news.com/ens/feb2003/2003-02-04-11.asp (accessed February 5, 2003).

Petersen, Melody. 2001. "German Company Tripling Production of Antibiotic Used to Fight Anthrax." *Santa Cruz Sentinel,* October 17, C-4.

Peterson, Jeannie, ed. 1983. *Nuclear War: The Aftermath.* Based on a special issue of *AMBIO* 11, 2–3 (1982). Oxford: Pergamon Press.

Peterson, Val. 1953. "Panic: The Ultimate Weapon?" *Collier's,* August 23, 99–109.

Pew Charitable Trust, Pew Initiative on Food and Biotechnology. 2005. "Public Sentiment about Genetically Modified Food." November, http://pewagbiotech.org/research/2005update/ (accessed May 10, 2006).

Pew Initiative on Food and Biotechnology. 2004. "Genetically Modified Crops in the United States 2004." www.mindfully.org/GE/2004/US-GMO-Crops-Pew1aug04.htm (accessed April 25, 2006).

Pew Research Center for the People and the Press. 2006. "Little Consensus on Global Warming; Partisanship Drives Opinion." July 12, people-press.org/reports/display.php3?ReportID=280 (accessed July 15, 2006).

Phillips, Kevin P. 1972. "The Emerging Republican Majority." In *The End of Innocence: A Suburban Reader,* ed. Charles M. Harr, 172–79. Glenview, Ill.: Scott, Foresman and Company.

Physicians for Social Responsibility. 2005. "A Brief Companion to CDC's 2005 National Exposure Report." www.envirohealthaction.org/upload_files/ACFEDA2.pdf (accessed April 14, 2006).

Piel, Gerard. 1962. "The Illusion of Civil Defense." *Bulletin of Atomic Scientists* 18, 2: 2–8.

Piketty, Thomas, and Emmanuel Saez. 2001. "Income Inequality in the United States. 1913–1998." NBER Working Paper No. 8467, September, www.nber.org/papers/w8467; series updated to 2000 at www .nber.org/data-appendix/w8467/TabFigs2000web.xls (accessed May 25, 2005).

Pollan, Michael. 2001. "Naturally: How Organic Became a Marketing Niche and a Multibillion-Dollar Industry." *New York Times Magazine,* May 13, 30–37, 57–58, 63–65.

———. 2006. *Omnivore's Dilemma.* New York: Penguin Press.

Pope, C. Arden, III, Richard T. Burnett, George D. Thurston, Michael J. Thun, Eugenia E. Calle, Daniel Krewski, and John J. Godleski. 2004. "Cardiovascular Mortality and Long-Term Exposure to Particulate Air Pollution: Epidemiological Evidence of General Pathophysiological Pathways of Disease." *Circulation* 109 (January 6/13): 71–77.

Porretto, John. 2003. "GM, Ford Enter Armored Car Market in Response to Growing Demand." *Detroit News,* March 29.

Postel, Sandra L., Gretchen C. Daily, and Paul R. Ehrlich. 1996. "Human Appropriation of Renewable Fresh Water." *Science* 271 (February 9): 785–88.

Pressler, Margaret Webb. 2003. "Growing, Naturally: Organic Foods Get into the Mainstream." *Washington Post,* July 26, E01.

Preston, Shelley. 2003. "As Demand Rises for Recycled and Chemical-Free Products, Designs and Styling Improve." *The Ledger,* Lakeland, Fla., December 13, search.theledger.com/apps/pbcs.dll/article?Date= 20031213&Category=NEWS&ArtNo=312130425&SectionCat=& Template=printart (accessed December 16, 2003).

Putnam, Robert D. 2000. *Bowling Alone: The Collapse and Revival of American Community.* New York: Simon & Schuster.

Ramankutty, Nevin, and Jonathan A. Foley. 1999. "Estimating Historical Changes in Global Land Cover: Croplands from 1700 to 1992." *Global Biochemical Cycles* 13, 4 (December): 997–1027.

Rauch, Molly, with P. W. McRandle. 2003. "A Fly in the Ointment: The Green Take on Insect Repellants and Sunscreens." *Grist Magazine,* June 18, www.gristmagazine.com/possessions/possessions061803 .asp?source=daily (accessed July 14, 2003).

Reeves, Richard. 1993. *President Kennedy: Profile of Power.* New York: Simon and Schuster.

Revkin, Andrew. 2002. "Federal Study Calls Spending on Water Systems Perilously Inadequate." *New York Times,* April 10, A22.

Rignot, Eric, and Pannir Kanagaratnam. 2006. "Changes in the Velocity Structure of the Greenland Ice Sheet." *Science* 311, 5763 (February 17): 986–90.

Riordan, Theresa. 2001. "Invention for a Jittery Public: Creating a Haven from Bioterrorism in the Living Room." *New York Times,* October 29, C2.

Roan, Shari. 2003. "Testing People for Pollutants: A Study Looking for Environmental Toxins in Breast-Milk Samples Puts California at the Forefront of the Biomonitoring Movement." *Los Angeles Times,* October 6, www.latimes.com/features/health/la-he-biomonitoring6oct06, 1,1246713.story (accessed October 7, 2003).

Rockefeller, Nelson A. 1960. "Importance of Shelters in Nuclear Age." *Vital Speeches of the Day,* April 15.

Rohter, Larry. 2002. "Punta Arenas Journal; In an Upside-Down World, Sunshine Is Shunned." *New York Times,* December 27, A4.

Root, Terry L., Jeff T. Price, Kimberly R. Hall, Stephen H. Schneider, Cynthia Rosenzweig, and J. Alan Pounds. 2003. "Fingerprints of Global Warming on Wild Animals and Plants." *Nature* 421 (January 2): 57–60.

Rose, Judy. 2001. "Americans Embracing Safe Room." *San Jose Mercury News,* June 23, 7G.

Rozhon, Tracie. 1999. "A Rockefeller Fixer-Upper." *New York Times,* Oct 14.

Ruben, Matthew. 2001. "Suburbanization and Urban Poverty under Neoliberalism." In *The New Poverty Studies: The Ethnography of Power, Politics, and Impoverished People in the United States,* ed. Judith Goode and Jeff Mashovsky, 435–69. New York: New York University Press.

Saad, Linda. 2006. "Americans Still Not Highly Concerned about Global Warming, Though Record Number Say It's Happening Now." Gallup Poll News Service, April 7, poll.gallup.com/content/default .aspx?ci=22291&VERSION=p (accessed July 29, 2006).

Sampson, Robert J., and William Julius Wilson. 1995. "Toward a Theory of Race, Crime and Urban Inequality." In *Crime and Inequality,* ed. John Hagan and Ruth D. Peterson, 37–54. Palo Alto, Calif.: Stanford University Press.

Sawyer, Scott. 2002. "Gemeinschaft or *Bust!* Suburban Sprawl, Environmental Degradation, and the New Urbanism Movement." Master's thesis, Department of Sociology, Washington State University.

Schafer, K. S., and S. E. Kegley. 2002. "Persistent Toxic Chemicals in the U.S. Food Supply." *Journal of Epidemiology and Community Health* 56, 11: 813–17.

Schafer, Kristin S., Margaret Reeves, Skip Spitzer, and Susan E. Kegley. 2004. "Chemical Trespass: Pesticides in Our Bodies and Corporate Accountability." Pesticide Action Network North America, San Francisco, www.panna.org/campaigns/docsTrespass/ChemTresMain(print) .pdf (accessed April 14, 2006).

Scheffer, Marten, Steve Carpenter, Jonathan A. Foley, and Brian Walker. 2001. "Catastrophic Shifts in Ecosystems." *Nature* 413 (October 11): 591–96.

Schell, Jonathan. 1988. *The Fate of the Earth.* New York: Knopf.

Schlesinger, Arthur M., Jr. 1965. *A Thousand Days: John F. Kennedy in the White House.* Boston: Houghton Mifflin.

Schmeltzer, John. 2005. "U.S. Develops Taste for Meat Seasoned with Sun, Fresh Air." *Chicago Tribune,* September 18.

Schneider, Conrad G. 2004. "Dirty Air, Dirty Power: Mortality and Health Damage Due to Air Pollution from Power Plants." Boston: Clean Air Task Force, June, www.cleartheair.org/dirtypower/docs/dirtyAir.pdf (accessed June 13, 2006).

Schneider, Conrad G., and L. Bruce Hill. 2005. "Diesel and Health in America: The Lingering Threat." Boston: Clean Air Task Force, February, www.catf.us/publications/reports/Diesel_Health_in_America.pdf (accessed June 13, 2006).

Schrope, Mark. 2000. "The Hole Story? Plankton Are Escaping the Ravages of Ozone Depletion—So Far." *New Scientist* 165, 2226 (February 19): 17.

Schupf, Harriet Warm. 1971. *The Perishing and Dangerous Classes: Efforts to Deal with the Neglected, Vagrant and Delinquent Juvenile in England, 1840–1875.* Ph.D. diss., Columbia University, University Microfilms, Ann Arbor, Mich.

Schwartz, Peter, and Doug Randall. 2003. "An Abrupt Climate Change Scenario and Its Implications for United States National Security." October, originally at www.ems.org/climate/pentagon_climatechange.pdf (accessed February 25, 2004); now available at www.environmentaldefense.org/documents/3566_AbruptClimateChange.pdf.

Sennett, Richard. 1970. *The Uses of Disorder: Personal Identity and City Life.* Harmondsworth, England: Penguin Books.

Severson, Kim. 2005. "An Organic Cash Cow." *New York Times,* December 9, D1, D10.

Sfiligoy, Eric. 1994. "Wait 'Til Next Year for Water Regs, Thanks to Congress's Aqua Dodge." *Beverage World's Periscope,* December 31, 1, 3.

Shapin, Steven. 2006. "Paradise Sold: What Are You Buying When You Buy Organic?" *New Yorker,* May 15, 84–88.

Shore, Allison. 1999. "Risk, Regulation, and Indoor Air Pollution: Environmental Inequality Inside." Master's thesis, Department of Sociology, University of California, Santa Cruz.

Shumsky, Neil L., ed. 1996. *American Cities: A Collection of Essays, Volume 1: Urbanization and the Growth of Cities.* New York: Garland Publishing.

Sidel, Victor W. 1966. "Medical Aspects of Civil Defense." In *Civil Defense,* ed. Henry Eyring, 53–75. Washington, D.C.: American Association for the Advancement of Science.

Simmons, Patrick A., and Robert E. Lang. 2003. "The Urban Turnaround." In *Redefining Urban and Suburban America: Evidence from Census 2000, Volume One,* ed. Bruce Katz and Robert E. Lang, 51–61. Washington, D.C.: Brookings Institution Press.

Simmons Market Research Bureau. 1992. "1992 Study of Media & Markets: Coffee, Tea, Cocoa, Milk, Soft Drinks, Juices & Bottled Water." P15. New York: Simmons Market Research Bureau.

Singer, J. David. 1961. "Deterrence and Shelters." *Bulletin of Atomic Scientist* 17, 8: 310–14.

Skonsen, Joel M. 1998. *Strategic Relocation: North American Guide to Safe Places.* American Fork, Utah: Swift Learning Resources.

Sohmer, Rebecca R., and Robert E. Lang. 2003. "Downtown Rebound." In *Redefining Urban and Suburban America: Evidence from Census 2000, Volume One,* ed. Bruce Katz and Robert E. Lang, 63–74. Washington, D.C.: Brookings Institution Press.

Sorensen, Theodore C. 1965. *Kennedy.* New York: Harper and Row.

Sorkin, Michael, ed. 1992. *Variations on a Theme Park: The New American City and the End of Public Space.* New York: Hill and Wang.

Spotts, Peter N. 2002. "What's in the Water? Better Detection Tools Reveal Possible Ecological 'Villains'—From Hormones to Fire Retardants—in U.S. Streams and Rivers." *Christian Science Monitor,* March 21, www.csmonitor.com/2002/0321/p11s02-sten.html (accessed March 21, 2002).

Squillace, Paul J., Michael J. Moran, Wayne W. Lapham, Curtis V. Price, Rick M. Clawges, and John S. Zogorski. 1999. "Volatile Organic Compounds in Untreated Ambient Groundwater of the United States, 1985–1995." *Environmental Science & Technology* 33, 23: 4176–87.

Stainforth, D. A, T. Aina, C. Christensen, M. Collins, N. Faull, D. J. Frame, J. A. Kettleborough, S. Knight, A. Martin, J. M. Murphy, C. Piani, D. Sexton, L. A. Smith, R. A. Spicer, A. J. Thorpe, and M. R. Allen. 2005. "Uncertainty in Predictions of the Climate Response to Rising Levels of Greenhouse Gases." *Nature* 433 (27 January): 403–6.

Stark, Andrew. 1998. "America, the Gated?" *Wilson Quarterly* 22, 1 (Winter): 58–79.

Stehlin, Dori. 1991. "Cosmetic Safety: More Complex Than at First Blush." U.S. Food and Drug Administration, *FDA Consumer,* November. www.cfsan.fda.gov/~dms/cos-safe.html (accessed February 5, 2005).

Steinemann, Anne. 2005. "Human Exposure and Health Hazards: Pt. 2." *Rachel's Environment & Health News* 811, March 31, www.rachel.org (accessed March 31, 2005).

Steinman, David, and Samuel S. Epstein. 1995. *The Safe Shopper's Bible: A Consumer's Guide to Nontoxic Household Products*. New York: Wiley.

Strauss, Gary. 2003. "Duct Tape Makers Swing into High Gear." *USA Today,* February 13, www.usatoday.com/money/industries/ manufacturing/2003-02-13-duct-tape_x.htm (accessed July 3, 2006).

Swanson, Sylvia E. 1995. "A Year in Review: Bottled Water Rises to Stardom." *Bottled Water Reporter,* December/January, 26–29.

Swayze, Jay. 1980. *Underground Gardens and Homes*. Hereford, Tex.: Geobuilding Systems.

Symons, Jelinger C. 1849. *Tactics for the Times: As Regards the Condition and Treatment of the Dangerous Classes*. London: John Ollivier, Pall Mall.

Szasz, Andrew. 1982. *The Dynamics of Social Regulation: A Study of the Formation and Evolution of the Occupational Safety and Health Administration*. Ph.D. diss., Department of Sociology, University of Wisconsin, Madison.

———. 1984. "Industrial Resistance toward Occupational Safety and Health Legislation, 1971–1981." *Social Problems* 32, 2: 103–16.

———. 1986a. "Corporations, Organized Crime and the Disposal of Hazardous Waste: An Examination of the Making of a Criminogenic Regulatory Structure." *Criminology* 24, 1: 1–27.

———. 1986b. "The Reversal of Federal Policy toward Worker Safety and Health: A Critical Examination of Alternative Explanations." *Science and Society* 50, 1: 25–51.

———. 1994. *EcoPopulism: Toxic Waste and the Movement for Environmental Justice*. Minneapolis: University of Minnesota Press.

Szasz, Andrew, and Michael Meuser. 1997. "Environmental Inequalities: Literature Review and Proposals for New Directions in Research and Theory." *Current Sociology* 45, 3: 99–120.

Tannen, Mary. 1994. "Eco-Yearnings." *New York Times Sunday Magazine,* March 20, 66.

Tarr, Joel L. 1996. *The Search for the Ultimate Sink: Urban Pollution in Historical Perspective*. Akron, Ohio: University of Akron Press.

Tavernise, Sabrina. 2004. "Watching Big Brother." *New York Times,* January 17, national ed., A14.

Thomas, Chris D., et al. 2004. "Extinction Risk from Climate Change." *Nature* 427 (January 8): 145–48.

Thomas, G. Scott. 1998. *The United States of Suburbia: How the Suburbs Took Control of America and What They Plan to Do with It*. Amherst, Mass.: Prometheus Books.

Thomas, William I., and Dorothy S. Thomas. 1928. *The Child in America*. New York: Alfred A. Knopf.

Thompson, Larry. 2000. "Are Bioengineered Foods Safe?" U.S. Food and Drug Administration, *FDA Consumer,* January-February. http://www.fda.gov/Fdac/features/2000/100_bio.html (accessed May 10, 2006).

"Trade Secrets: A Moyers Report." 2001. Directed by Joseph Camp. Produced by Public Affairs Television in association with Washington Media Associates. Originally aired on PBS March 26, 2001.

Trent, Martin, and Jerry Felice. 1990. "1990 Bottled Water Survey." Bureau of Drinking Water, Suffolk County Department of Health Services, May.

Tucker, Carol. 1998. "Gated Communities: The Barriers Go Up." *Public Management* 80, 5 (May): 22–24.

United Nations Environment Programme. 1998. "Environmental Effects of Ozone Depletion, 1998 Assessment; Executive Summary." *Journal of Photochemistry and Photobiology B: Biology* 46: 1–4. www.gcrio.org/ozone/execsum.pdf. (accessed July 2, 2006).

U.S. Census Bureau. 2000. 2000 Census, www.census.gov/hhes/www/income/income00/inctab4.html (accessed September 1, 2005).

U.S. Congress. House. Committee on Energy and Commerce, Subcommittee on Oversight and Investigations. 1991a. *Bottled Water Regulation: Hearing before the Subcommittee on Oversight and Investigations of the Committee on Energy and Commerce.* 102nd Cong., 1st sess., April 10, 1991. Washington, D.C.: U.S. Government Printing Office.

———. House. Committee on Energy and Commerce, Subcommittee on Oversight and Investigations. 1991b. *Proceedings of the Bottled Water Workshop (September 13 and 14, 1990).* Washington, D.C.: U.S. Government Printing Office.

———. Joint Committee on Atomic Energy. 1959. *Biological and Environmental Effects of Nuclear War. Hearings before the Special Subcommittee on Radiation of the Joint Committee on Atomic Energy.* 86th Cong., 1st sess. Washington, D.C.: Government Printing Office.

U.S. Department of Agriculture (USDA). 2000. Food Safety and Inspection Service, 2000 FSIS National Residue Program, the "Bluebook." "Appendix III: U.S. Residue Limits for Pesticides in Meat, Poultry, and Egg Products." www.fsis.usda.gov/OPHS/blue2000/appendix3.pdf (accessed March 9, 2002).

U.S. Department of Defense, Office of Civil Defense. 1968. *Radiological Defense: Textbook.* SM-11.22-2, June. Washington, D.C.: Department of Defense.

U.S. Department of Health and Human Services. 2001. "First National Report on Human Exposure to Environmental Chemicals." National Center for Environmental Health, Centers for Disease Control and Prevention, March, NCEH Pub. No. 01-0379, Atlanta.

————. 2003. "Second National Report on Human Exposure to Environmental Chemicals." National Center for Environmental Health, Centers for Disease Control and Prevention, January, http://www.cdc.gov/exposurereport%3ENCEH Pub. No. 03-0022, Atlanta.

————. 2005. "Third National Report on Human Exposure to Environmental Chemicals." National Center for Environmental Health, Centers for Disease Control and Prevention, July, NCEH Pub. No. 05-0570, Atlanta.

U.S. Environmental Protection Agency (U.S. EPA). 1997a. "Health and Environmental Effects of Ground-Level Ozone." Office of Air and Radiation, Office of Air Quality Planning and Standards, July 17, www.epa.gov/ttn/oarpg/naaqsfin/o3health.html (accessed May 29, 2003).

————. 1997b. "Health and Environmental Effects of Particulate Matter." Office of Air and Radiation, Office of Air Quality Planning and Standards, July 17, www.epa.gov/ttn/oarpg/naaqsfin/pmhealth.html (accessed May 29, 2003).

————. 1999a. "Understanding the Safe Drinking Water Act." EPA 810-F-99-008, December.

————. 1999b. "Twenty-five Years of the Safe Drinking Water Act: Protecting Our Health from Source to Tap." EPA 810-K-99-004: 3, December.

————. 1999c. "National-Scale Air Toxics Assessment for 1999: Estimated Emissions, Concentrations and Risk, Technical Fact Sheet." Technology Transfer Network, http://www.epa.gov/ttn/atw/nata1999/natafinalfact.html (accessed May 26, 2006).

————. 2000a. "National Water Quality Inventory: 1998 Report to Congress." EPA841-R-00-001, October.

————. 2000b. "Atlas of America's Polluted Waters." EPA 840-B-00-002.

————. 2000c. "The Safe Drinking Water Act Amendments of 1996." www.epa.gov/safewater/sdwa/theme.html (accessed July 23, 2001).

————. 2000d. "Estimated Per Capita Water Ingestion in the United States: Based on Data Collected by the United States Department of Agriculture's 1994–96 Continuing Survey of Food Intakes by Individuals." Office of Water, EPA-822-R-00-008, April.

————. 2000e. "National Air Quality and Emissions Trends Report, 1998." Office of Air Quality Planning and Standards, Emissions Monitoring and Analysis Division, Air Quality Trends Analysis Group, 454/R-00-003, Research Triangle Park, N.C., March.

————. 2001a. "Toxics Release Inventory 1999 Executive Summary." EPA 260-R-01-001, May 2.

————. 2001b. "Drinking Water Contaminant Candidate List." July 23, http://epa.gov/ogwdw/ccl/cclfs.html (accessed November 18, 2001).

————. 2001c. "1999 Drinking Water Infrastructure Needs Survey." Office of Water, EPA 816-R-01-004, February.

————. 2002. "The Clean Water and Drinking Water Infrastructure Gap Analysis." Office of Water, EPA-816-R-02-020, September.

————. 2003a. "National Air Quality and Emissions Trends Report, 2003 Special Studies Edition." Office of Air Quality Planning and Standards, Emissions Monitoring and Analysis Division, Air Quality Trends Analysis Group September, Research Triangle Park, N.C., www.epa.gov/air/airtrends/aqtrnd03/ (accessed June 10, 2006).

————. 2003b. "Analysis and Findings of The Gallup Organization's Drinking Water Customer Satisfaction Survey." EPA 816-K-03-005, August.

————. 2005a. "Drinking Water Contaminant Candidate List (CCL)." http://www.epa.gov/safewater/ccl/frequentquestions.html (accessed July 21, 2006).

————. 2005b. "Unregulated Contaminant Monitoring Rule 2 (UCMR 2)." http://www.epa.gov/safewater/ucmr/ucmr2/ (accessed July 21, 2006).

————. 2006a. "2004 TRI Public Data Release eReport, Data Tables and Charts, Section B—Summary." April, www.epa.gov/tri/tridata/tri04/pdfs/Section_B_03-14-06.pdf (accessed July 23, 2006).

————. 2006b. "Air Emissions Trends: Continued Progress through 2005." April 20, www.epa.gov/airtrends/2006/econ-emissions.html (accessed May 19, 2006).

U.S. Environmental Protection Agency and the U.S. Consumer Product Safety Commission (CPSC). 1995. "The Inside Story: A Guide to Indoor Air Quality." Office of Radiation and Indoor Air, EPA Document # 402-K-93-007, April, www.epa.gov/egi-bin/epaprintonly.cgi (accessed May 29, 2003).

U.S. Federal Civil Defense Administration. n.d. "FACTS about Fallout." Washington, D.C.: Government Printing Office.

————. 195?. "Bert the Turtle Says Duck and Cover." Reprinted by Office of Civil Defense, State of California, Sacramento.

————. 1950. "Survival under Atomic Attack." NSRB Doc. 130, Washington, D.C.: Government Printing Office.

————. 1953. "Home Shelters for Family Protection in an Atomic Attack." Technical Manual 5-5. Washington, D.C.: Government Printing Office.

U.S. Food and Drug Administration. 2003. "Distribution of Dioxin-Contaminated Animal Feed Mineral Mixes and Feed Halted." FDA Talk Paper, T03-17, February 28, www.fda.gov/bbs/topics/ANSWERS/2003/ANS01203.html (accessed May 12, 2006).

U.S. General Accounting Office (U.S. GAO). 1991a. "Drinking Water: Inadequate Regulation of Home Treatment Units Leaves Consumers at Risk." GAO/RCED-92-34, Washington, D.C., December.

———. 1991b. "Food Safety and Quality: Stronger FDA Standards and Oversight Needed for Bottled Water." GAO/RCED-91-67, Washington, D.C., March.

———. 1999a. "Safe Drinking Water Act: Progress and Future Challenges in Implementing the 1996 Amendments." GAO/RCED-99-31, January.

———. 1999b. "Animal Agriculture: Waste Management Practices." RCED-99-205, June.

———. 1999c. "Drinking Water Research: Better Planning Needed to Link Needs and Resources." GAO/RCED-99-273, September.

———. 2000. "Water Efficiency Standards." Washington, D.C.: Government Printing Office, GAO/RCED-00-161R.

———. 2006. "Clean Air Act: EPA Should Improve the Management of Its Air Toxics Program." GAO-06-669, June, www.gao.gov/new.items/d06669.pdf (accessed July 27, 2006).

U.S. Geological Survey (USGS). 1999. "The Quality of Our Nation's Waters; Nutrients and Pesticides." U.S. Geological Survey Circular 1225, http://water.usgs.gov/pubs/circ/circ1225/pdf/index.html.

Van Eenennaam, Alison L. 2005. "Genetic Engineering and Animal Feed." University of California, Davis, Division of Agriculture and Natural Resources, Genetic Engineering Fact Sheet 6, Publication 8183, anrcatalog.ucdavis.edu/pdf/8183.pdf (accessed May 15, 2006).

van Heugten, Eric. 1997. "Water Quality." *Swine News,* November, North Carolina Cooperative Extension Service, mark.asci.ncsu.edu/Swine_News/1997/sn_v2010.htm (accessed April 21, 2006).

Veblen, Thorstein. 1899. *The Theory of the Leisure Class: An Economic Study of Institutions.* New York: Macmillan Company.

Vincent, Roger. 2005. "Organic Beauty Products Get a Lift with USDA About-Face." *Los Angeles Times,* August 25.

Vitousek, Peter M., Paul R. Ehrlich, Anne H. Ehrlich, and Pamela A. Matson. 1986. "Human Appropriation of the Products of Photosynthesis." *BioScience* 36, 6 (June): 368–73.

Vitousek, Peter M., Harold A. Mooney, Jane Lubchenco, and Jerry M. Melillo. 1997. "Human Domination of Earth's Ecosystems." *Science* 277 (July 25): 494–99.

von Wiesenberger, Arthur. 1999. "Bottled Waters of the Rich and Famous." www.bottledwaterweb.com/articles/avw-0003.htm (accessed March 23, 2001).

Waldman, Peter. 2005. "Common Industrial Chemicals in Tiny Doses Raise Health Issue." *Wall Street Journal,* July 25, 1.

Warner, Melanie. 2005. "What Is Organic? Powerful Players Want a Say." *New York Times,* November 1, C1, C4.

Waskow, Arthur I., and Stanley L. Newman. 1962. *America in Hiding.* New York: Ballantine Books.

Watson, Bruce. 1984. "We Couldn't Run, So We Hoped We Could Hide." *Smithsonian* 25, 1 (April): 46–57.

Weart, Spencer R. 1987. "History of American Attitudes to Civil Defense." In *Civil Defense: A Choice of Disasters,* ed. John Dowling and Evans M. Harrell, 11–32. New York: American Institute of Physics.

———. 1988. *Nuclear Fear: A History of Images.* Cambridge, Mass.: Harvard University Press.

———. 2003a. "The Discovery of Global Warming." www.aip.org/history/climate/rapid.htm (accessed August 28, 2003).

———. 2003b. "The Discovery of Rapid Climate Change." *Physics Today* 56, 8 (August): 30–36.

Weisel, Clifford P., and Wan-Kuen Jo. 1996. "Ingestion, Inhalation, and Dermal Exposures to Chloroform and Trichloroethene from Tap Water." *Environmental Health Perspectives* 104, 1 (January): 48–51.

Weisel, Clifford P., Hekap Kim, Patricia Haltmeier, and Judith B. Klotz. 1999. "Exposure Estimates to Disinfection By-Products of Chlorinated Drinking Water." *Environmental Health Perspectives* 107, 2 (February): 103–10.

Weiss, Harvey, and Raymond S. Bradley 2001. "What Drives Societal Collapse?" *Science* 291, 5504 (January 26): 609–10.

Whole Foods Market. n.d. "Whole Foods Market Nationwide Survey Reveals More Than Half of Americans Have Sampled Organic Foods and That Food Labels Matter." www.wholefoodsmarket.com/company/pr_organicsurvey.pdf (faxed to author March 14, 2003).

Wigner, Eugene P. 1966. "The Possible Effectiveness of Civil Defense." In *Civil Defense,* ed. Henry Eyring, 33–51. Washington, D.C.: American Association for the Advancement of Science.

———, ed. 1968. *Who Speaks for Civil Defense?* New York: Charles Scribner's Sons.

Williams, Robert C., and Philip L. Cantelon, eds. 1984. *The American Atom: A Documentary History of Nuclear Policies from the Discovery of Fission to the Present, 1939–1984.* Philadelphia: University of Pennsylvania Press.

Wilson, William Julius. 1987. *The Truly Disadvantaged: The Inner City, the Underclass, and Public Policy.* Chicago: University of Chicago Press.

Winkler, Alan M. 1984. "A Forty-Year History of Civil Defense." *Bulletin of the Atomic Scientists* 40 (June/July): 16–22.

Witchel, Alex. 2000. "Water Pressure: Never Higher." *New York Times,* June 14, B1, B12.

World Wildlife Fund. 2004. "Bad Blood? A Survey of Chemicals in the Blood of European Ministers." October, worldwildlife.org/toxics/pubs/badblood.pdf (accessed April 13, 2006).

Yang, Sarah. 2005. "Transgenic Plants Remove More Selenium from Contaminated Soil than Wild-type Plants, New Field Tests Show." UC Berkeley News Press Release, February 1, www.berkeley.edu/news/media/releases/2005/02/01_plantremediation.shtml (accessed April 21, 2006).

Yankelovich Partners. 2000. "Beverage Consumption in 2000—Final Report." Prepared for the International Bottled Water Association, March.

Yess, Norma J., Ellis L. Gunderson, and Ronald R. Roy. 1993. "U.S. Food and Drug Administration Monitoring of Pesticide Residues in Infant Foods and Adult Foods Eaten by Infants/Children." *Journal of the Association of Official Analytical Chemists International* 76, 3 (May-June): 492–507.

Yoon, Carol Kaesuk. 2003. "Exposure to Pesticides Is Lowered When Young Children Go Organic." *New York Times*, March 25, D2.

Index

Andrew Szasz is professor and chair of the sociology department at the University of California, Santa Cruz. He is the author of *EcoPopulism: Toxic Waste and the Movement for Environmental Justice* (Minnesota, 1994).